T0074268

Springer Theses

Recognizing Outstanding Ph.D. Research

For further volumes:
http://www.springer.com/series/8790

Aims and Scope

The series "Springer Theses" brings together a selection of the very best Ph.D. theses from around the world and across the physical sciences. Nominated and endorsed by two recognized specialists, each published volume has been selected for its scientific excellence and the high impact of its contents for the pertinent field of research. For greater accessibility to non-specialists, the published versions include an extended introduction, as well as a foreword by the student's supervisor explaining the special relevance of the work for the field. As a whole, the series will provide a valuable resource both for newcomers to the research fields described, and for other scientists seeking detailed background information on special questions. Finally, it provides an accredited documentation of the valuable contributions made by today's younger generation of scientists.

Theses are accepted into the series by invited nomination only and must fulfill all of the following criteria

- They must be written in good English.
- The topic should fall within the confines of Chemistry, Physics, Earth Sciences, Engineering and related interdisciplinary fields such as Materials, Nanoscience, Chemical Engineering, Complex Systems and Biophysics.
- The work reported in the thesis must represent a significant scientific advance.
- If the thesis includes previously published material, permission to reproduce this must be gained from the respective copyright holder.
- They must have been examined and passed during the 12 months prior to nomination.
- Each thesis should include a foreword by the supervisor outlining the significance of its content.
- The theses should have a clearly defined structure including an introduction accessible to scientists not expert in that particular field.

Tsukasa Mizuhara

Development of Novel Anti-HIV Pyrimidobenzothiazine Derivatives

Doctoral Thesis accepted by
Kyoto University, Japan

 Springer

Author (Current address)
Dr. Tsukasa Mizuhara
University of Massachusetts
Amherst, MA
USA

Supervisor
Prof. Nobutaka Fujii
Kyoto University
Kyoto
Japan

ISSN 2190-5053   ISSN 2190-5061 (electronic)
ISBN 978-4-431-54444-9   ISBN 978-4-431-54445-6 (eBook)
DOI 10.1007/978-4-431-54445-6
Springer Tokyo Heidelberg New York Dordrecht London

Library of Congress Control Number: 2013941348

Printed on acid-free paper

Springer is part of Springer Science+Business Media (www.springer.com)

Parts of this thesis have been published in the following journal articles:

Tsukasa Mizuhara, Shinsuke Inuki, Shinya Oishi, Nobutaka Fujii, and Hiroaki Ohno, "Cu(II)-Mediated Oxidative Intermolecular Ortho C–H Functionalisation Using Tetrahydropyrimidine as the Directing Group", Chemical Communications, 2009, (23), 3413–3415.

Tsukasa Mizuhara, Shinya Oishi, Nobutaka Fujii, and Hiroaki Ohno, "Efficient Synthesis of Pyrimido[1,2-c][1,3]benzothiazin-6-imines and Related Tricyclic Heterocycles by S_NAr-Type C–S, C–N, or C–O Bond Formation with Heterocumulenes", The Journal of Organic Chemistry, 2010, 75 (1), 265–268.

Tsukasa Mizuhara, Shinya Oishi, Hiroaki Ohno, Kazuya Shimura, Masao Matsuoka, and Nobutaka Fujii, "Concise Synthesis and Anti-HIV Activity of Pyrimido[1,2-c][1,3]benzothiazin-6-imines and Related Tricyclic Heterocycles", Organic & Biomolecular Chemistry, 2012, 10 (33), 6792–6802.

Tsukasa Mizuhara, Shinya Oishi, Hiroaki Ohno, Kazuya Shimura, Masao Matsuoka, and Nobutaka Fujii, "Structure–Activity Relationship Study of Pyrimido[1,2-c][1,3]benzothiazin-6-imine Derivatives for Potent Anti-HIV Agents", Bioorganic & Medicinal Chemistry, 2012, 20 (21), 6334–6441.

Tsukasa Mizuhara, Shinya Oishi, Hiroaki Ohno, Kazuya Shimura, Masao Matsuoka, and Nobutaka Fujii, "Design and Synthesis of Biotin- or Alkyne-Conjugated Photoaffinity Probes for Studying the Target Molecules of PD 404182", Bioorganic & Medicinal Chemistry, 2013, 21 (7), 2079–2087.

Supervisor's Foreword

It is my pleasure to introduce Dr. Tsukasa Mizuhara's thesis for the Springer Theses Prize, as an outstanding original work in one of the world's top universities. He joined my group at Kyoto University as a graduate student in April 2008 and started his doctoral study under my supervision and that of Drs. Hiroaki Ohno and Shinya Oishi. In March 2013, his doctoral thesis was accepted by Kyoto University and he obtained his Ph.D. degree.

Dr. Mizuhara initiated his study by screening novel anti-HIV compounds in collaboration with Prof. Masao Matsuoka (Institute of Virus Research, Kyoto University) and identified several candidates including PD 404182, which originally had been reported as an antibacterial agent. He focused on PD 404182-lead optimization and developed two novel synthetic methods to facilitate the SAR study of PD 404182 with unique tricyclic pyrimido[1,2-c][1,3]benzo-thiazin-6-imine structure. One is reported in "Cu(II)-Mediated Oxidative Intermolecular *Ortho* C–H Functionalisation Using Tetrahydropyrimidine as the Directing Group" (*Chemical Communications*, 2009, Jun 21;(23):3413–3415) and the other in "Efficient Synthesis of Pyrimido[1,2-c][1,3]benzothiazin-6-imines and Related Tricyclic Heterocycles by S_NAr-Type C–S, C–N, or C–O Bond Formation with Heterocumulenes" (*The Journal of Organic* Chemistry, 2010, Jan 1;75(1): 265–268).

Having the efficient synthetic methods in hand, he carried out intensive SAR studies of the central 1,3-thiazin-2-imine core, the benzene part, and the cyclic amidine part of PD 404182 for the development of more effective anti-HIV agents. The 6-6-6 fused pyrimido[1,2-c][1,3]benzothiazine scaffold and the heteroatom arrangement in PD 404182 contribute considerably to the potent anti-HIV activity. Additionally, through optimization studies of the benzene and the cyclic amidine ring parts in PD 404182, three-fold more potent inhibitors were identified. He also revealed by a time-of-drug-addition experiment that PD 404182 derivatives impaired HIV replication at an early stage of the infection cycle (around the binding or fusion stage). Finally, he identified eight possible protein targets by using photoaffinity probes designed through SAR study of PD 404182.

This thesis is very comprehensive (including synthetic organic chemistry, medicinal chemistry, virology, and chemical biology) and informative for developing novel antivirus agents since PD 404182 is reported to be effective even against hepatitis C virus (HCV) as well as HIV-1 and HIV-2. It is noteworthy that all of this work was based on Dr. Mizuhara's exceptionally original ideas. Five outstanding papers related to this thesis, prepared by himself as the first author, have been published in the top journals of organic chemistry and medicinal chemistry.

Kyoto, Japan, April 26, 2013   Nobutaka Fujii

Acknowledgments

I would like to express my sincere and wholehearted appreciation to Professor Nobutaka Fujii (Graduate School of Pharmaceutical Sciences, Kyoto University) for his kind guidance, constructive discussions, and constant encouragement during this study. Thanks also go to Dr. Hiroaki Ohno (Graduate School of Pharmaceutical Sciences, Kyoto University) and Dr. Shinya Oishi (Graduate School of Pharmaceutical Sciences, Kyoto University) for their valuable suggestions, guidance, and support throughout this study. The support and advice of Professor Masao Matsuoka (Institute for Virus Research, Kyoto University), Dr. Kazuya Shimura (Institute for Virus Research, Kyoto University), Professor Hideaki Kakeya (Graduate School of Pharmaceutical Sciences, Kyoto University), Professor Yoshiji Takemoto (Graduate School of Pharmaceutical Sciences, Kyoto University), and Professor Kiyosei Takasu (Graduate School of Pharmaceutical Sciences, Kyoto University) were greatly appreciated.

I wish to express my gratitude to Professor Munetaka Kunishima (Faculty of Pharmaceutical Sciences, Institute of Medical, Pharmaceutical, and Health Sciences, Kanazawa University) and Dr. Kazuhito Hioki (Faculty of Pharmaceutical Sciences, Kobe Gakuin University).

I am indebted to Dr. Hideki Maeta, Dr. Masahiko Taniguchi, Mr. Takayuki Kato, Ms. Kumiko Hiyama, Mr. Shuhei Osaka, Dr. Megumi Okubo, Dr. Daisuke Nakagawa, Mr. Tatsuya Murakami, Dr. Kazunobu Takahashi, and Dr. Hideki Kurihara for excellent technical assistance.

I also wish to express my gratitude to Dr. Shinsuke Inuki and all the other colleagues in the Department of Bioorganic Medicinal Chemistry/Department of Chemogenomics (Graduate School of Pharmaceutical Sciences, Kyoto University) for their valuable comments and for their assistance and cooperation in various experiments.

I thank the Japan Society for the Promotion of Science (JSPS) for financial support, Dr. Naoshige Akimoto for mass spectral measurements, and all the staff at the Elemental Analysis Center, Kyoto University.

Finally, I thank my parents, Takashi and Kumiko Mizuhara, my sister, Chie Moriwaki, my brother, Masaru Mizuhara, and my brother-in-low, Masaki Moriwaki, for their understanding and constant encouragement through this study.

Contents

Chapter 1
Introduction

Human immunodeficiency virus (HIV) infection remains one of the most serious threats to public health. HIV is a lentivirus that causes acquired immunodeficiency syndrome (AIDS) [1], in which progressive failure of the immune system allows life-threatening opportunistic infections. According to estimates by the UNAIDS Report 2012, about 34 million people worldwide are living with HIV; approximately 2.5 million people are newly infected with HIV; and more than 25 million patients have died of AIDS [2]. This global health threat has triggered intensive drug discovery efforts and a number of anti-HIV drugs including azidothymidine [AZT; the first nucleoside reverse transcriptase inhibitor (NRTI)], saquinavir (the first protease inhibitor), and nevirapine [a non-nucleoside reverse transcriptase inhibitor (NNRTI)] has been approved for treatment of HIV infection (Fig. 1.1) [3]. Highly active antiretroviral therapy (HAART) involving co-administration of these anti-HIV agents is a standard treatment regimen for HIV infection. This regimen suppresses HIV replication and controls disease progression in HIV-infected patients [4, 5]. Unfortunately, however, an increasing number of patients with HIV infection/AIDS have failed to respond to the current antiretroviral therapeutics because of serious problems including the emergence of drug-resistant HIV variants [6] and drug-related adverse effects [7]. To overcome these problems, several antiretroviral agents with new mechanisms of action have been developed in this decade (Fig. 1.1). A peptide-based fusion inhibitor (enfuvirtide) [8–10], an integrase inhibitor (raltegravir) [11], and a CC chemokine receptor type 5 (CCR5) antagonist (maraviroc) [12, 13] are examples of new molecular entities used as anti-HIV agents.

Recently, highly potent small-molecule anti-HIV agents have been reported, which inhibit the binding or fusion of HIV to host cells (Fig. 1.2). 2-Thioxo-1,3-thiazolidine derivative **1** shows potent inhibition of HIV-1 replication at the nanomolar level [14, 15], which is directed at the deep hydrophobic pocket in the N-terminal heptad repeat trimer of the viral gp41. Compound **1** blocks HIV-1-mediated cell–cell fusion and the formation of gp41 six-helix bundles, as does enfuvirtide [15]. Small-molecule CD4 mimics with oxalamide and related substructures are another series of anti-HIV agents [16–18]. The representative BMS-448043 (**2**) exhibits subnanomolar anti-HIV activity by interaction with the CD4

T. Mizuhara, *Development of Novel Anti-HIV Pyrimidobenzothiazine Derivatives*,
Springer Theses, DOI: 10.1007/978-4-431-54445-6_1, © Springer Japan 2013

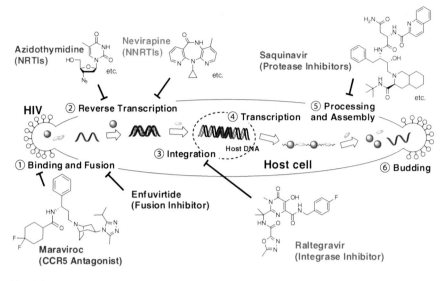

Fig. 1.1 Structures of approved anti-HIV agents and the target process in HIV-1 life cycle

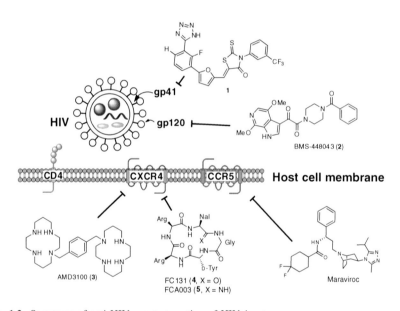

Fig. 1.2 Structures of anti-HIV agents targeting of HIV-1 entry process

binding pocket in gp120 [18]. Alternatively, there are some examples of antagonists against CXC chemokine receptor type 4 (CXCR4). AMD 3100 (**3**) exhibits potent anti-CXCR4 activity [19, 20]. Crystal structure analysis of receptor and mutation experiments have indicated that the ion-pairing interaction between the basic functional groups of the antagonist and acidic residues in CXCR4 contributes to the potent activity [21, 22]. Fujii et al. have reported highly potent CXCR4 antagonists [23]. FC131 (**4**) was identified from a library of cyclic pentapeptides consisting of pharmacophore residues of the polyphemusin-II-derived anti-HIV peptide T140 [24]. FCA003 (**5**) is an amidine isostere-containing analog of FC131 [25]. Analysis of the FC131–CXCR4 complex model has revealed that these cyclic peptides effectively bind with CXCR4 to inhibit the interaction of CXCR4 with gp120 [26–29]. These entry inhibitors will provide alternative combination regimens of anti-HIV agents for the treatment of drug-resistant variants.

To develop the novel antiretroviral small molecules, random screening of small molecules using multinuclear activation of a galactosidase indicator (MAGI) assay has been carried out [30], in which the inhibitory activity of early stage HIV infection, including virus attachment and membrane fusion to host cells, is evaluated. Among >30,000 compounds screened, 3,4-dihydro-2H,6H-pyrimido[1,2-c][1, 3]benzothiazin-6-imine (**6**; PD 404182, Fig. 1.3) was identified as a potent anti-HIV agent lead.

Compound **6** has previously been reported to have inhibitory activity against 3-deoxy-D-manno-octulosonic acid 8-phosphate synthase [31] and phosphopantetheinyl transferase [32, 33], exerting antimicrobial effects. Additionally, in the course of this study, antiviral activities of **6** against hepatitis C virus (HCV) and simian immunodeficiency virus (SIV) were reported [34, 35]. Dual infection with HIV and HCV confers a higher rate of viral resistance and higher susceptibility to death compared with solely HIV-infected patients [36]. Approximately 5 million HIV-patients are co-infected with HCV; therefore, antiviral agents against multiple viruses are favorable.

Intensive investigations of the structure–activity relationship (SAR) and mechanistic studies have been carried out for lead optimization of **6**. This thesis consists of three chapters: (1) the development of efficient synthetic methods of PD 404182 and related tricyclic heterocycles by sp^2-carbon–heteroatom bond formation; (2) SAR study of PD 404182 for development of potent anti-HIV agents; and (3) synthesis and application of photoaffinity probes of PD 404182 for target identification.

Section 2.1 describes a Cu-mediated oxidative intermolecular *ortho* C–H functionalization using tetrahydropyrimidine as the directing group. Using several nucleophiles, C–O, C–N, and C–S bonds were formed by this reaction, which facilitated synthesis of various tricyclic heterocycles.

Section 2.2 presents a direct synthesis of tricyclic heterocycles by a regioselective S_NAr-type reaction of tetrahydropyrimidine-substituted haloarenes with heterocumulenes. The efficient synthetic route for pyrimido[1,2-c] [1, 3] benzothiazin-6-imine such as PD 404182 is also described.

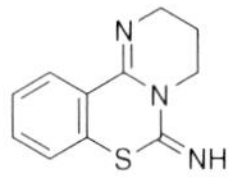

PD 404182 (**6**)
$EC_{50} = 0.44 \pm 0.08$ μM
$CC_{50} > 100$ μM

Fig. 1.3 Structure of PD 404182

Chapter 3 describes the SAR study of PD 404182 derivatives. The structural elements to give the highly potent anti-HIV agents are revealed. In addition, the mechanism of action of PD 404182 derivatives is discussed.

Chapter 4 describes the design and synthesis of biotin- or alkyne-conjugated photoaffinity probes of PD 404182 and their application to the target identification experiment for HIV-1-infected H9 cells.

References

1. Weiss, R.A.: Science **260**, 1273–1279 (1993)
2. UNAIDS Report on the Global AIDS Epidemic: http://www.unaids.org/en/media/unaids/contentassets/documents/epidemiology/2012/gr2012/20121120_UNAIDS_Global_Report_2012_en.pdf (2012)
3. Esté, J.A., Cihlar, T.: Antivi. Res. **85**, 25–33 (2010)
4. Thompson, M.A., Aberg, J.A., Cahn, P., Montaner, J.S., Rizzardini, G., Telenti, A., Gatell, J.M., Günthard, H.F., Hammer, S.M., Hirsch, M.S., Jacobsen, D.M., Reiss, P., Richman, D.D., Volberding, P.A., Yeni, P., Schooley, R.T.J.: Am. Med. Assoc. **304**, 321–333 (2010)
5. Rathbun, R.C., Lockhart, S.M., Stephens, J.R.: Curr. Pharm. Des. **12**, 1045–1063 (2006)
6. Johnson, V.A., Calvez, V., Günthard, H.F., Paredes, R., Pillay, D., Shafer, R., Weinsing, A.M., Richman, D.D.: Top. Antivir. Med. **19**, 156–164 (2011)
7. Carr, A., Cooper, D.A.: Lancet **356**, 1423–1430 (2000)
8. Kilby, J.M., Eron, J.J.N.: Engl. J. Med. **348**, 2228–2238 (2003)
9. Lalezari, J.P., Henry, K., O'Hearn, M., Montaner, J.S., Piliero, P.J., Trottier, B., Walmsley, S., Cohen, C., Kuritzkes, D.R., Eron Jr, J.J., Chung, J., DeMasi, R., Donatacci, L., Drobnes, C., Delehanty, J., Salgo, M.N.: Engl. J. Med. **348**, 2175–2185 (2003)
10. Matthews, T., Salgo, M., Greenberg, M., Chung, J., DeMasi, R., Bolognesi, D.: Nat. Rev. Drug Discov. **3**, 215–225 (2004)
11. Steigbigel, R.T., Cooper, D.A., Kumar, P.N., Eron, J.E., Schechter, M., Markowitz, M., Loutfy, M.R., Lennox, J.L., Gatell, J.M., Rockstroh, J.K., Katlama, C., Yeni, P., Lazzarin, A., Clotet, B., Zhao, J., Chen, J., Ryan, D.M., Rhodes, R.R., Killar, J.A., Gilde, L.R., Strohmaier, K.M., Meibohm, A.R., Miller, M.D., Hazuda, D.J., Nessly, M.L., DiNubile, M.J., Isaacs, R.D., Nguyen, B.-Y., Teppler, H.N.: Engl. J. Med. **359**, 339–354 (2008)
12. Dorr, P., Westby, M., Dobbs, S., Griffin, P., Irvine, B., Macartney, M., Mori, J., Rickett, G., Smith-Burchnell, C., Napier, C., Webster, R., Armour, D., Price, D., Stammen, B., Wood, A., Perros, M.: Antimicrob. Agents Chemother. **49**, 4721–4732 (2005)
13. Fätkenheuer, G., Pozniak, A.L., Johnson, M.A., Plettenberg, A., Staszewski, S., Hoepelman, A.I.M., Saag, M.S., Goebel, F.D., Rockstroh, J.K., Dezube, B.J., Jenkins, T.M., Medhurst, C., Sullivan, J.F., Ridgway, C., Abel, S., James, I.T., Youle, M., Van Der Ryst, E.: Nat. Med. **11**, 1170–1172 (2005)
14. Katritzky, A.R., Tala, S.R., Lu, H., Vakulenko, A.V., Chen, Q.-Y., Sivapackiam, J., Pandya, K., Jiang, S., Debnath, A.K.J.: Med. Chem. **52**, 7631–7639 (2009)
15. Jiang, S., Tala, S.R., Lu, H., Abo-Dya, N.E., Avan, I., Gyanda, K., Lu, L., Katritzky, A.R., Debnath, A.K.J.: Med. Chem. **54**, 572–579 (2011)
16. Lin, P.-F., Blair, W.S., Wang, T., Spicer, T.P., Guo, Q., Zhou, N., Gong, Y.-F., Wang, H.-G.H., Rose, R., Yamanaka, G., Robinson, B., Li, C.-B., Fridell, R., Deminie, C., Demers, G., Yang, Z., Zadjura, L., Meanwell, N.A., Colonno, R.J.: Proc. Natl. Acad. Sci. U.S.A. **100**, 11013–11018 (2003)

17. Si, Z., Madani, N., Cox, J.M., Chruma, J.J., Klein, J.C., Schon, A., Phan, N., Wang, W., Biorn, A.C., Cocklin, S., Chaiken, I., Freire, E., Smith, A.B., Sodroski, J.G.: Proc. Natl. Acad. Sci. U.S.A. **101**, 5036–5041 (2004)
18. Wang, T., Yin, Z., Zhang, Z., Bender, J.A., Yang, Z., Johnson, G., Yang, Z., Zadjura, L.M., D'Arienzo, C.J., Parker, D.D., Gesenberg, C., Yamanaka, G.A., Gong, Y.-F., Ho, H.-T., Fang, H., Zhou, N., McAuliffe, B.V., Eggers, B.J., Fan, L., Nowicka-Sans, B., Dicker, I.B., Gao, Q., Colonno, R.J., Lin, P.-F., Meanwell, N.A., Kadow, J.F.J.: Med. Chem. **52**, 7778–7787 (2009)
19. Bridger, G.J., Skerlj, R.T., Thornton, D., Padmanabhan, S., Martellucci, S.A., Henson, G.W., Abrams, M.J., Yamamoto, N., De Vreese, K., Pauwels, R., De Clercq, E.J.: Med. Chem. **38**, 7366–7378 (1995)
20. Donzella, G.A., Schols, D., Lin, S.W., Esté, J.A., Nagashima, K.A., Maddon, P.J., Allaway, G.P., Sakmar, T.P., Henson, G., De Clercq, E., Moore, J.P.: Nat. Med. **4**, 772–777 (1998)
21. Rosenkilde, M.M., Gerlach, L.-O., Jakobsen, J.S., Skerlj, R.T., Bridger, G.J., Schwartz, T.W.J.: Biol. Chem. **279**, 3033–3041 (2004)
22. Wu, B., Chien, E.Y.T., Mol, C.D., Fenalti, G., Liu, W., Katritch, V., Abagyan, R., Brooun, A., Wells, P., Bi, F.C., Hamel, D.J., Kuhn, P., Handel, T.M., Cherezov, V., Stevens, R.C.: Science **330**, 1066–1071 (2010)
23. Oishi, S., Fujii, N.: Org. Biomol. Chem. **10**, 5720–5731 (2012)
24. Fujii, N., Oishi, S., Hiramatsu, K., Araki, T., Ueda, S., Tamamura, H., Otaka, A., Kusano, S., Terakubo, S., Nakashima, H., Broach, J.A., Trent, J.O., Wang, Z., Peiper, S.C.: Angew. Chem. Int. Ed. **42**, 3251–3253 (2003)
25. Inokuchi, E., Oishi, S., Kubo, T., Ohno, H., Shimura, K., Matsuoka, M., Fujii, N.: ACS Med. Chem. Lett. **2**, 477–480 (2011)
26. Våbenø, J., Nikiforovich, G.V., Marshall, G.R.: Chem. Biol. Drug Des. **67**, 346–354 (2006)
27. Våbenø, J., Nikiforovich, G.V., Marshall, G.R.: Biopolymers **84**, 459–471 (2006)
28. Demmer, O., Dijkgraaf, I., Schumacher, U., Marinelli, L., Cosconati, S., Gourni, E., Wester, H.-J., Kessler, H.J.: Med. Chem. **54**, 7648–7662 (2011)
29. Yoshikawa, Y., Kobayashi, K., Oishi, S., Fujii, N., Furuya, T.: Bioorg. Med. Chem. Lett. **22**, 2146–2150 (2012)
30. Watanabe, K., Negi, S., Sugiura, Y., Kiriyama, A., Honbo, A., Iga, K., Kodama, E.N., Naitoh, T., Matsuoka, M., Kano, K.: Chem. Asian J. **5**, 825–834 (2010)
31. Birck, M.R., Holler, T.P., Woodard, R.W.J.: Am. Chem. Soc. **122**, 9334–9335 (2000)
32. Duckworth, B.P., Aldrich, C.C.: Anal. Biochem. **403**, 13–19 (2010)
33. Foley, T.L., Yasgar, A., Garcia, C.J., Jadhav, A., Simeonov, A., Burkart, M.D.: Org. Biomol. Chem. **8**, 4601–4606 (2010)
34. Chockalingam, K., Simeon, R.L., Rice, C.M., Chen, Z.: Proc. Natl. Acad. Sci. U.S.A. **107**, 3764–3769 (2010)
35. Chamoun, A.M., Chockalingam, K., Bobardt, M., Simeon, R., Chang, J., Gallay, P., Chen, Z.: Antimicrob. Agents Chemother. **56**, 672–681 (2012)
36. Operskalski, E.A., Kovacs, A.: Curr. HIV/AIDS Rep. **8**, 12–22 (2011)

Chapter 2
Development of Divergent Synthetic Methods of Pyrimidobenzothiazine and Related Tricyclic Heterocycles

2.1 Cu(II)-Mediated *Ortho*-Selective Intermolecular C–H Functionalization

3,4-Dihydro-2*H*,6*H*-pyrimido[1,2-*c*][1,3]benzothiazin-6-imine (PD 404182, **1**, Fig. 2.1) is a promising anti-HIV agent lead discovered by a random screening project. To develop the highly potent derivatives, it is valuable to establish practical and short-step synthetic approaches for the preparation of several derivatives[1,2]. The author planned to develop a diversity-oriented approach to synthesize tricyclic heterocycles related to PD 404182 based on the sp^2-carbon–heteroatom (O, N, and S) bond formations (Scheme 2.1). It was expected that the *ortho*-selective introduction of a heteroatom on 2-phenyl-1,4,5,6-tetrahydropyrimidine derivatives **3** [6], which is easily obtained from the corresponding benzaldehydes **2**, followed by functional group transformations leads to various types of heterocycles **5** including PD 404182.

Directing group-assisted intermolecular C–H functionalization is considered to be one of the most promising approaches for constructions of various heterocycles, providing several biologically active compounds since a new or carbon–heteroatom bond is selectively formed at a non-functionalized position proximal to the directing group. (Scheme 2.2)[3,4]. In general, C–H functionalization proceeds via metallacycle formation by oxidative addition of transition-metal and subsequent coordination of nucleophile and reductive elimination. Recent research has revealed that nitrogen-containing functional groups such as pyridines [14–16],

---

[1] In the previous reports, compound **1** was obtained via benzo-1,2-dithiole-3-thiones and 2-(1,4,5,6-tetrahydro-2-pyrimidinyl)benzenethiol in 3 % yield from 2-chlorobenzyl chloride, see [1–3]

[2] See [4–5].
[3] For reviews on transition-metal-catalyzed directed C–H activations, See [7–9].
[4] See [10–13].

T. Mizuhara, *Development of Novel Anti-HIV Pyrimidobenzothiazine Derivatives*, Springer Theses, DOI: 10.1007/978-4-431-54445-6_2, © Springer Japan 2013

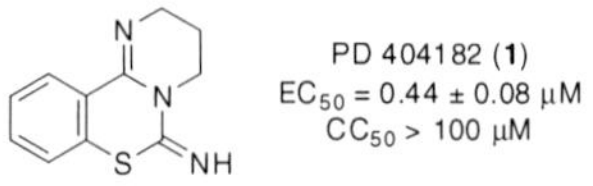

PD 404182 (1)
$EC_{50} = 0.44 \pm 0.08\ \mu M$
$CC_{50} > 100\ \mu M$

Fig. 2.1 Structure of PD 404182

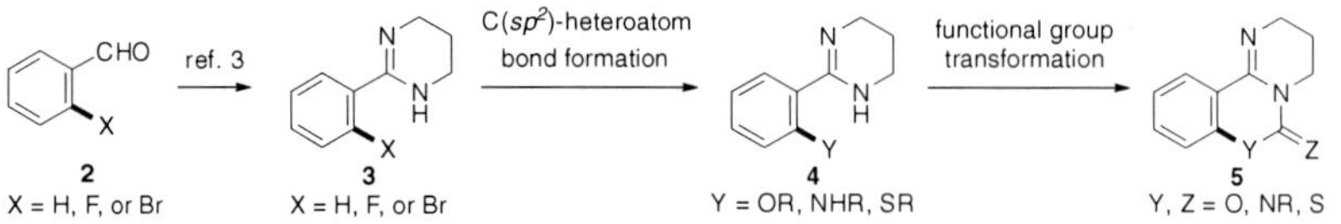

Scheme 2.1 Synthetic scheme for PD 404182 derivatives via carbon–heteroatom bond formation

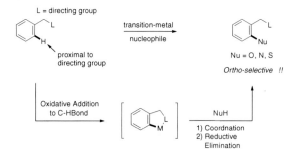

Scheme 2.2 Carbon–heteroatom bond formation by C–H functionalization

imines [17–19], oxazolines [20, 21], and amidines [22] effectively act as directing groups for regioselective C–H functionalization.

Cu-mediated reactions[5] have facilitated the synthesis of biologically active compounds because of its cost, earth abundance, and lower toxicity. Reinaud and co-workers have reported a Cu-mediated *ortho*-hydroxylation reaction of benzamide **6** using a carboxyl group as a directing group (Scheme 2.3, eq 1) [24]. Yu et al. (eq 2) [25] and Chatani et al. (eq 3) [26] have independently reported Cu-mediated oxidative intermolecular C–H functionalization using a pyridine moiety as a directing group. The author designed an experiment for the oxidative introduction of heteroatoms by aromatic C–H functionalization with the assistance of an *ortho*-tetrahydropyrimidinyl group (eq 4).

A few recent reports have revealed that amidine moieties effectively act as directing groups for the *ortho*-selective C–H functionalization (Scheme 2.4). Inoue and co-workers have reported *ortho*-selective arylation of 2-arylimidazolines with aryl halides in the presence of a Ru(II)–phosphine complex [22]. The reaction of

---

[5] For a review on Cu-mediated C–H functionalization, see [23].

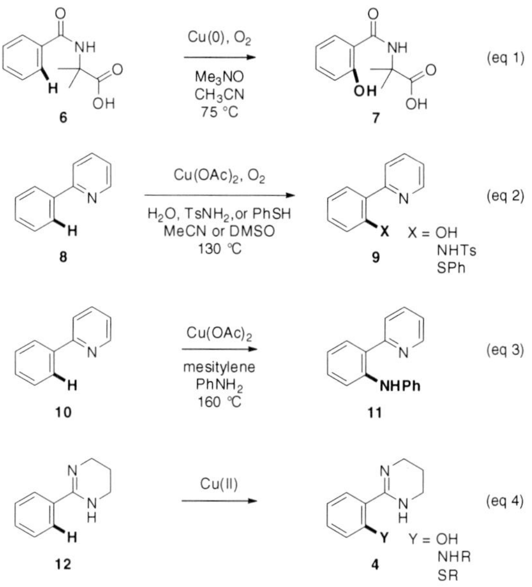

Scheme 2.3 Cu-mediated intermolecular C–H functionalization

2-phenylimidazoline **13** with 1.2 equiv of bromobenzene using $[RuCl_2(\eta^6\text{-}C_6H_6)]_2$ yielded the mono- and diarylated products (**14** and **15**) in a 64 % yield and in 31:69 ratio (eq 1). Buchwald and co-workers have reported the formation of aryl-benz-imidazole **17** by $Cu(OAc)_2$-catalyzed oxidative cyclization of amidine **16** [27]. The best result was obtained by using 15 mol % of $Cu(OAc)_2$ and 2–5 equiv of HOAc under an O_2 atmosphere. In this reaction, an amidine moiety acts as a directing group as well as a nucleophile. These contributions prompted the author to investigate a tetrahydropyrimidine group-assisted regioselective C–H functionalization.

The author initially investigated the reaction conditions for C–H hydroxylation (Table 2.1). In the presence of H_2O (1.0 equiv), treatment of 2-phenyl-1,4,5,6-tetrahydropyrimidine (**18a**) with CuO, $Cu(OH)_2$, $Cu(OTf)_2$ or $Cu(tfa)_2$ (1.0 equiv) in DMF at 130 °C under an O_2 atmosphere led to the recovery of unchanged starting material and the desired C–H oxidation did not occur (entries 1–4). Using $Cu(OAc)_2$, [25, 26] however, led to the formation of the desired *ortho*-hydroxyl-ated compound **19a** (ca. 69 % yield) although the product yield of compound **19a** was poorly reproducible because of its high basicity. The author then attempted to isolate **20a** as the tricyclic PD 404182 derivative: after the disappearance of **18a** (monitored by TLC), the solvent was evaporated *in vacuo* and the treatment with triphosgene (1.05 equiv) and triethylamine (4.0 equiv) in CH_2Cl_2 afforded pure **20a** in a yield of 61 % (entry 5, Condition A). When acetonitrile or dioxane was used as the solvent instead of DMF, yields of **20a** decreased considerably (11 %, entries 6 and 7). Lowering the loading of $Cu(OAc)_2$ to 0.2 equiv also resulted in a decreased yield for **20a** (30 %, entry 8), which indicates low catalyst efficiency.

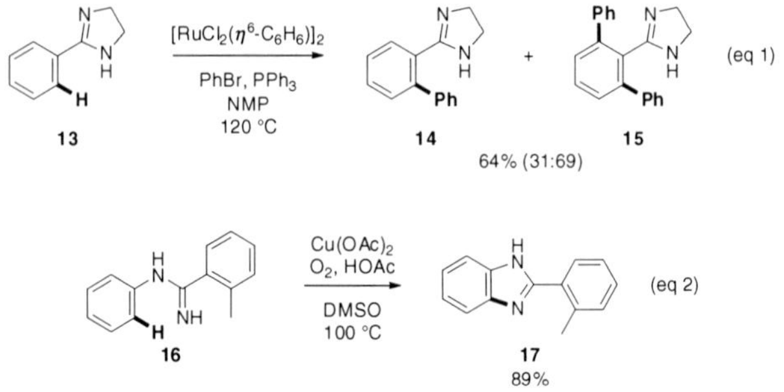

Scheme 2.4 Amidine directed regioselective C–H functionalization

Table 2.1 Optimization of reaction conditions for C–H hydroxylation[a]

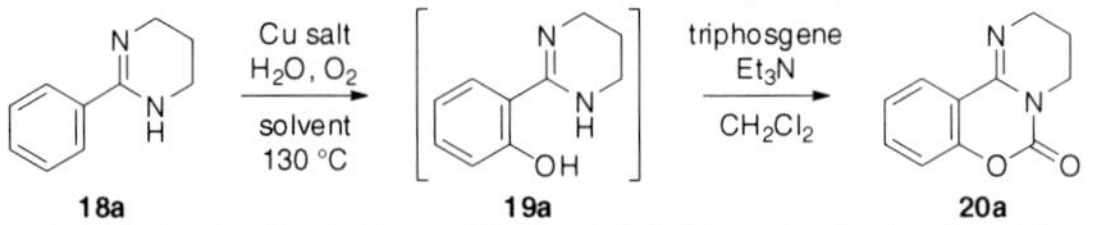

Entry	Cu salt (equiv)	Solvent	Time (min)	Yield (%)[b]
1	CuO (1.0)	DMF	20	No reaction
2	Cu(OH)$_2$ (1.0)	DMF	20	No reaction
3	Cu(OTf)$_2$ (1.0)	DMF	20	No reaction
4	Cu(tfa)$_2$ (1.0)	DMF	20	No reaction
5	Cu(OAc)$_2$ (1.0)	DMF	20	61
6	Cu(OAc)$_2$ (1.0)	MeCN	60	11
7	Cu(OAc)$_2$ (1.0)	Dioxane	60	11
8	Cu(OAc)$_2$ (1.0)	DMF	60	30
9	Cu(OAc)$_2$ (2.0)	DMF	15	27
10[c]	Cu(OAc)$_2$ (1.0)	DMF	20	70
11[c,d]	Cu(OAc)$_2$ (1.0)	DMF	20	56

[a] After completion of C–H hydroxylation (monitored by TLC), the reaction mixture was evaporated and treated with triphosgene (1.05 equiv) and Et$_3$N (4.0 equiv) in CH$_2$Cl$_2$ at 0 °C to rt for 1 h (Condition A)
[b] Isolated yields
[c] After completion of C–H hydroxylation (monitored by TLC), the reaction mixture was treated with TMEDA (4.0 equiv) at 130 °C for 1 min. In this case, TMEDA (additional 4.0 equiv) was used for the next step instead of Et$_3$N (Condition B)
[d] Reaction was carried out under air
TMEDA = N,N,N',N'-tetramethylethylenediamine

When using 2.0 equiv of Cu(OAc)$_2$, the yield also decreased contrary to the author's expectation (27 %, entry 9).

Considering that the *ortho*-hydroxylated product **19a** may form a complex with the Cu salt, the author further optimized the reaction conditions including the carbonylation procedure. Initially, N,N,N',N'-tetramethylethylenediamine (TMEDA) was added as a bidentate ligand to the oxidative C–H functionalization reaction mixture and this resulted in the complete inhibition of the desired transformation. Similarly, use of TMEDA instead of triethylamine as a base for the carbonylation did not improve the yield of **20a**. On the other hand, treatment with TMEDA (4.0 equiv) at 130 °C for 1 min after the C–H hydroxylation followed by the carbonylation using additional TMEDA (4.0 equiv) increased the yield to 70 % (entry 10, Condition B). The reaction under air resulted in a decreased yield (56 %, entry 11).

Using the condition B, the author examined the reaction of several substituted substrates (Table 2.2). Substitution with electron-donating groups such as methoxy (**18b**, entry 1) or methyl groups (**18c**, entry 2) was tolerated to afford the desired products **20b** and **20c** in 64 % and 61 % yields, respectively. The chemoselectivity of this reaction was evaluated by a reaction where aryl bromide **18d** was used and the desired product **20d** was obtained in a 45 % yield (entry 3). Methoxycarbonyl (entry 4) and trifluoromethyl groups (entry 5) had relatively small effects on the reactivity of these substrates and the use of the highly electron-deficient arene **18g** bearing a nitro group decreased the yield considerably (19 %, entry 6). These results indicate that this reaction is sensitive to the presence of electron-withdrawing groups on the aromatic ring. In all cases, reactions under condition A gave less favorable results.

To confirm the actual source of *ortho*-hydroxyl group, the author carried out the C–H hydroxylation reaction using $H_2^{18}O$ under an Ar atmosphere (Scheme 2.5). This reaction provided compound **20a** with ^{16}O, suggesting that *ortho*-hydroxyl group was derived from Cu(OAc)$_2$ [25] Notably, the reaction under an Ar atmosphere gave the product in low yield, suggesting that molecular O$_2$ participates in the reoxidation of the Cu catalyst.

Next, the author investigated the ability of other amidine analogues to function as directing groups (Fig. 2.2). The reaction of the *N*-methylated analog **21** and 2-phenylimidazole **22** did not produce the desired *ortho*-hydroxylated products under the standard reaction conditions and the starting materials were recovered. Unexpectedly, the five-membered ring amidine in **13** was not effective as a directing group either. These results suggest that subtle differences in the intermediate formed by a Cu salt and a directing group strongly affect the reactivity of the substrates.

Although the exact mechanism of the *ortho* C–H oxidation is unclear, on the basis of these observations and the seminal work of others, the author proposes the two possible reaction mechanisms (Scheme 2.6): a single electron transfer (SET) pathway (A) [25, 28] and electrophilic substitution pathway (B) [27]. In pathway A, Cu–N adduct **II** is initially formed by the reaction of compound **I** with Cu(OAc)$_2$ [29, 30]. A SET from an aryl ring to the coordinated Cu(II) led to radical cation intermediate **III**. Intramolecular acetate transfer followed by another SET step and transfer of a proton yielded the acetoxylated compound **V**. Subsequent hydrolysis gave an *ortho*-hydroxylated product **VI**. The observed *ortho*-selectivity could be attributed to an intramolecular transfer of the coordinating group on the Cu atom. Alternatively, in pathway B, addition of a π-system to the Cu center yielded metallacycle **VII**. Compound **V** was formed through

Table 2.2 Cu-mediated C–H hydroxylation of *Para*-substituted-2-phenyl-1,4,5,6-tetrahydro-pyrimidines[a]

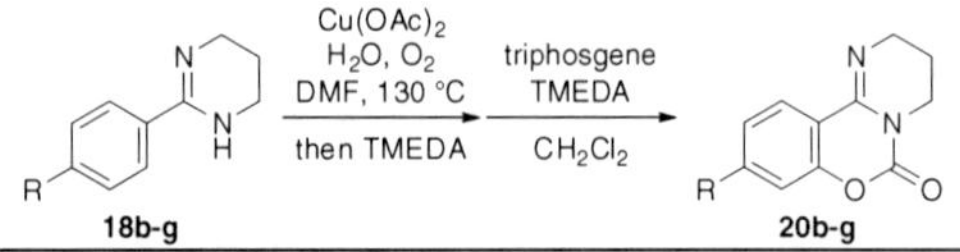

18b-g 20b-g

Entry	Substrate (R)	Product	Yield (%)[b]
1	**18b** (OMe)	20b	64 (53)
2	**18c** (Me)	20c	61 (54)
3	**18d** (Br)	20d	45 (37)
4	**18e** (CO_2Me)	20e	46 (43)
5	**18f** (CF_3)	20f	43 (38)
6	**18g** (NO_2)	20g	19 (16)

[a] These reactions were carried out using the optimized procedure (Condition B, Table 2.1, entry 10)
[b] Isolated yields. Yields in parentheses indicate those of the reactions at condition A (Table 2.1, entry 5)

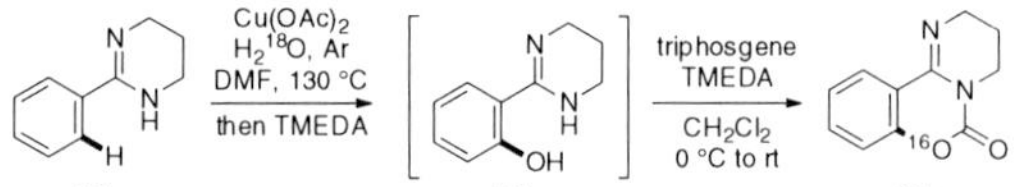

18a 19a 20a
17%

Scheme 2.5 C–H hydroxylation reaction with $H_2^{18}O$

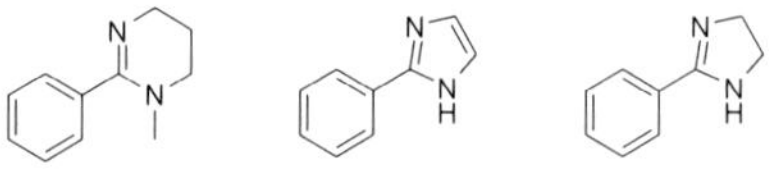

21 22 13

Fig. 2.2 Various amidine analogs used for the *ortho*-hydroxylation experiments

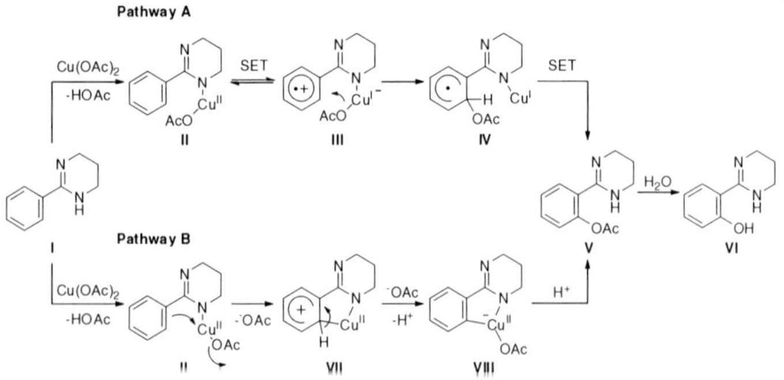

Scheme 2.6 Proposed reaction mechanisms

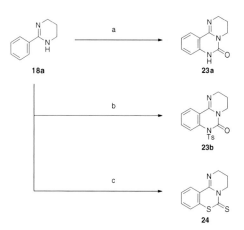

Scheme 2.7 C–N and C–S bond formations with various nucleophiles. *Reagents and conditions:* (a) Cu(OAc)$_2$, BocNH$_2$, O$_2$, DMF, 100 °C, 53 %; (b) (i) Cu(OAc)$_2$, TsNH$_2$, O$_2$, DMF, 130 °C; (ii) triphosgene, Et$_3$N, CH$_2$Cl$_2$, 0 °C to rt, 47 % (2 steps); (c) Cu(OAc)$_2$, CS$_2$, O$_2$, 1,4-dioxane, 130 °C, 11 %

rearomatization and subsequent reductive elimination. These mechanisms are supported by the findings that the presence of an electron-withdrawing group on the benzene ring considerably decreased the product yields. Recently, involvement of Cu(III) species in the C–H oxidation reaction has been demonstrated.[6,7] Therefore, it is possible that this reaction proceeded via the formation of Cu(III)–substrate **I** complex.

Finally, the author investigated C–N and C–S bond formations (Scheme 2.7). The author found that the reaction of amidine **18a** with Cu(OAc)$_2$ (1.0 equiv) and *tert*-butyl carbamate (3.0 equiv) in DMF at 100 °C for 40 min directly afforded the tricyclic aniline derivative (**23a**) in 53 % yield. This reaction occurred by cyclization involving the elimination of *tert*-butoxide. *p*-Toluenesulfonamide [37] also reacted with **18a** under identical condition to afford **23b** in 47 % yield after alumina column chromatography[8] followed by treatment with triphosgene–Et$_3$N. In addition, the reaction with CS$_2$ in 1,4-dioxane at 130 °C directly gave pyrimido[1,2-*c*][1,3]benzothiazine derivative **24**. The C–N and C–S bond forming

---

[6] For a review on high-valent Cu(III) species in catalysis, see [31].

[7] See [32–36].

[8] Because the separation of **23b** and the by-product **20a** was difficult, separation by alumina column chromatography was necessary before carbonylation.

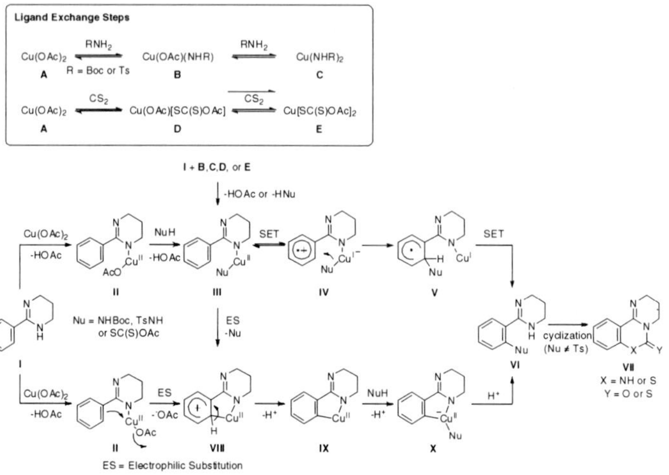

Scheme 2.8 Proposed reaction mechanisms of C–N and C–S bond formations

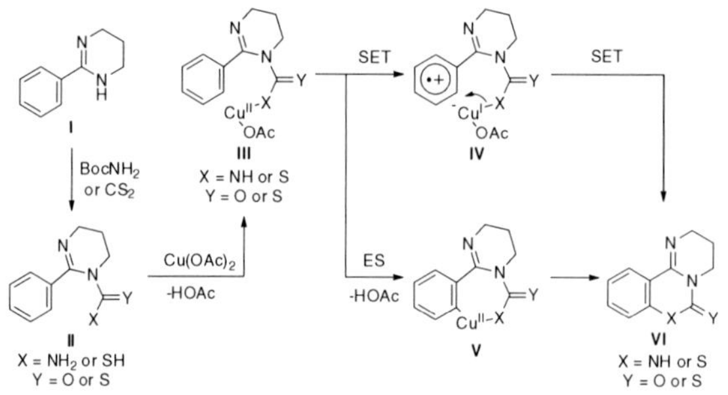

Scheme 2.9 Alternative proposed reaction mechanisms with BocNH$_2$ and CS$_2$

reactions can be explained by a similar mechanism as depicted in Scheme 2.7 including ligand exchange step.[9,10]

In conclusion, the author has developed a Cu-mediated oxidative *ortho* C–H functionalization using tetrahydropyrimidine as a directing group. This reaction was applied to 2-phenyl-1,4,5,6-tetrahydropyrimidines having an electron-donating or a weak electron-withdrawing group to afford the corresponding pyrimido[1,2-c][1,3]benzoxazine derivatives. Use of *tert*-butyl carbamate, *p*-toluenesulfonamide,

---

[9] Examples for the reaction mechanism including the ligand exchange are shown above (Scheme 2.8)

[10] Different reaction pathways are not excluded at present. Some examples are shown above (Scheme 2.9).

or CS_2 instead of H_2O promotes the introduction of a nitrogen or sulfur functionality to give pyrimido[1,2-c]quinazoline or pyrimido[1,2-c][1,3]benzothiazine derivative, respectively.

2.1.1 Experimental Section

2.1.1.1 General Methods

All moisture-sensitive reactions were performed using syringe-septum cap techniques under an Ar atmosphere and all glasswares were dried in an oven at 80 °C for 2 h prior to use. Melting points were measured by a hot stage melting point apparatus (uncorrected). For flash chromatography, Wakogel C-300E (Wako) or aluminum oxide 90 standardized (Merck) was employed. ^{1}H-NMR spectra were recorded using a JEOL AL-400 or a JEOL ECA-500 spectrometer, and chemical shifts are reported in δ (ppm) relative to Me_4Si ($CDCl_3$) as internal standards. ^{13}C-NMR spectra were recorded using a JEOL AL-400 or JEOL ECA-500 spectrometer and referenced to the residual $CHCl_3$ signal. ^{19}F-NMR spectra were recorded using a JEOL ECA-500 and referenced to the internal $CFCl_3$ (δ_F 0.00 ppm). ^{1}H-NMR spectra are tabulated as follows: chemical shift, multiplicity (b = broad, s = singlet, d = doublet, t = triplet, q = quartet, m = multiplet), coupling constant(s), and number of protons. Exact mass (HRMS) spectra were recorded on a JMS-HX/HX 110A mass spectrometer. Infrared (IR) spectra were obtained on a JASCO FT/IR-4100 FT-IR spectrometer with JASCO ATR PRO410-S.

2.1.1.2 General Procedure for Preparation of the Substrates. Synthesis of 2-Phenyl-1,4,5,6-tetrahydropyrimidine (18a)

To a solution of benzaldehyde (5.00 g, 47.1 mmol) in *t*-BuOH (470 mL) was added propylenediamine (3.84 g, 51.8 mmol). After being stirred at 70 °C for 30 min, K_2CO_3 (19.53 g, 141.3 mmol) and I_2 (14.95 g, 58.8 mmol) were added. After being stirred at same temperature for 3 h, the reaction mixture was quenched with sat. Na_2SO_3 until the iodine color disappeared. The organic layer was separated and concentrated *in vacuo*. The resulting solid was dissolved with H_2O, and then pH was adjusted to 12–14 with 2 N NaOH. The whole was extracted with $CHCl_3$ and dried over $MgSO_4$. After concentration, the resulting solid was recrystallized from $CHCl_3$–*n*-hexane to give the title compound 18a as colorless crystals (6.62 g, 82 %): mp 88–89 °C (from $CHCl_3$–*n*-hexane); IR (neat) cm^{-1}: 1618 (C=N); ^{1}H-NMR (400 MHz, $CDCl_3$) δ: 1.83–1.85 (m, 2H, CH_2), 3.49 (t, J = 5.9 Hz, 4H, 2 × CH_2), 5.02 (br s, 1H, NH), 7.34–7.38 (m, 3H, Ar), 7.63–7.66 (m, 2H, Ar); ^{13}C-NMR (100 MHz, $CDCl_3$) δ: 20.7, 42.3 (2C), 126.0 (2C), 128.2

(2C), 129.6, 137.3, 154.5; *Anal.* calcd for $C_{10}H_{12}N_2$: C, 74.97; H, 7.55; N, 17.48. Found; C, 74.79; H, 7.53; N, 17.43.

2.1.1.3 2-(4-Methoxyphenyl)-1,4,5,6-tetrahydropyrimidine (18b)

p-Methoxybenzaldehyde (1.36 g, 10 mmol) was subjected to the general procedure. Colorless crystals (1.40 g, 74 %): mp 132–134 °C (from $CHCl_3$–*n*-hexane); IR (neat) cm^{-1}: 1611 (C=N); ^{1}H-NMR (400 MHz, $CDCl_3$) δ: 1.81–1.87 (m, 2H, CH_2), 3.49 (t, *J* = 5.7 Hz, 4H, 2 × CH_2), 3.81 (s, 3H, OCH_3), 4.87 (br s, 1H, NH), 6.86 (d, *J* = 9.4 Hz, 2H, Ar), 7.60 (d, *J* = 9.4 Hz, 2H, Ar); ^{13}C-NMR (100 MHz, $CDCl_3$) δ: 20.9, 42.4 (2C), 55.2, 113.5 (2C), 127.2 (2C), 130.0, 153.9, 160.6; *Anal.* calcd for $C_{11}H_{14}N_2O$: C, 69.45; H, 7.42; N, 14.73. Found: C, 69.18; H, 7.46; N, 14.58.

2.1.1.4 2-(4-Tolyl)-1,4,5,6-tetrahydropyrimidine (18c)

p-Tolualdehyde (1.20 g, 10 mmol) was subjected to the general procedure. Colorless crystals (1.03 g, 59 %): mp 120–121 °C (from $CHCl_3$–*n*-hexane); IR (neat) cm^{-1}: 1615 (C=N); ^{1}H-NMR (400 MHz, $CDCl_3$) δ: 1.82–1.85 (m, 2H, CH_2), 2.35 (s, 3H, CH_3), 3.49 (t, *J* = 5.7 Hz, 4H, 2 × CH_2), 4.90 (br s, 1H, NH), 7.15 (d, *J* = 8.3 Hz, 2H, Ar), 7.54 (d, *J* = 8.3 Hz, 2H, Ar); ^{13}C-NMR (100 MHz, $CDCl_3$) δ: 20.8, 21.2, 42.3 (2C), 125.8 (2C), 128.9 (2C), 134.5, 139.5, 154.3; *Anal.* calcd for $C_{11}H_{14}N_2$: C, 75.82; H, 8.10; N, 16.08. Found: C, 75.76; H, 8.01; N, 15.91.

2.1.1.5 2-(4-Bromophenyl)-1,4,5,6-tetrahydropyrimidine (18d)

p-Bromobenzaldehyde (1.85 g, 10 mmol) was subjected to the general procedure. Colorless crystals (1.82 g, 76 %): mp 174–175 °C (from $CHCl_3$–*n*-hexane); IR (neat) cm^{-1}: 1619 (C=N); ^{1}H-NMR (400 MHz, $CDCl_3$) δ: 1.81–1.88 (m, 2H, CH_2), 3.49 (t, *J* = 5.7 Hz, 4H, 2 × CH_2), 4.81 (br s, 1H, NH), 7.48 (d, *J* = 8.8 Hz, 2H, Ar), 7.53 (d, *J* = 8.8 Hz, 2H, Ar); ^{13}C-NMR (100 MHz, $CDCl_3$) δ: 20.7, 42.4 (2C), 123.8, 127.6 (2C), 131.4 (2C), 136.3, 153.5; *Anal.* calcd for $C_{10}H_{11}BrN_2$: C, 50.23; H, 4.64; N, 11.72. Found: C, 50.20; H, 4.51; N, 11.66.

2.1.1.6 Methyl 4-(1,4,5,6-tetrahydropyrimidin-2-yl)benzoate (18e)

Methyl 4-formylbenzoate (1.00 g, 6.09 mmol) was subjected to the general procedure. Colorless crystals (1.63 g, 80 %): mp 152–153 °C (from $CHCl_3$–*n*-hexane); IR (neat) cm^{-1}: 1721 (C=O), 1620 (C=N); ^{1}H-NMR (400 MHz, $CDCl_3$) δ:

1.83–1.89 (m, 2H, CH$_2$), 3.52 (t, J = 5.7 Hz, 4H, 2 × CH$_2$), 3.92 (s, 3H, OCH$_3$), 5.04 (br s, 1H, NH), 7.72 (d, J = 8.5 Hz, 2H, Ar), 8.02 (d, J = 8.5 Hz, 2H, Ar); ^{13}C-NMR (100 MHz, CDCl$_3$) δ: 20.5, 42.3 (2C), 52.1, 126.0 (2C), 129.5 (2C), 130.8, 141.5, 153.6, 166.6; *Anal.* calcd for C$_{12}$H$_{14}$N$_2$O$_2$: C, 66.04; H, 6.47; N, 12.84. Found: C, 65.76; H, 6.28; N, 12.69.

2.1.1.7 2-[4-(Trifluoromethyl)phenyl]-1,4,5,6-tetrahydropyrimidine (18f)

p-(Trifluoromethyl)benzaldehyde (1.74 g, 10 mmol) was subjected to the general procedure. Colorless crystals (1.71 g, 75 %): mp 176–177 °C (from CHCl$_3$–*n*-hexane); IR (neat) cm^{-1}: 1620 (C=N); ^{1}H-NMR (400 MHz, CDCl$_3$) δ: 1.83–1.89 (m, 2H, CH$_2$), 3.51 (t, J = 5.7 Hz, 4H, 2 × CH$_2$), 4.92 (br s, 1H, NH), 7.61 (d, J = 8.3 Hz, 2H, Ar), 7.76 (d, J = 8.3 Hz, 2H, Ar); ^{13}C-NMR (100 MHz, CDCl$_3$) δ: 20.6, 42.4 (2C), 122.6, 125.2 (q, J = 3.7 Hz, 2C), 126.4 (2C), 131.4 (d, J = 32.3 Hz), 140.7, 153.3; ^{19}F-NMR (500 MHz, CDCl$_3$) δ: −62.6; *Anal.* calcd for C$_{11}$H$_{11}$F$_3$N$_2$: C, 57.89; H, 4.86; N, 12.28. Found: C, 57.89; H, 4.82; N, 12.29.

2.1.1.8 2-(4-Nitrophenyl)-1,4,5,6-tetrahydropyrimidine (18g)

p-Nitrobenzaldehyde (1.51 g, 10 mmol) was subjected to the general procedure. Yellow crystals (1.63 g, 80 %): mp 169–171 °C (from CHCl$_3$–*n*-hexane); IR (neat) cm^{-1}: 1623 (C=N), 1519 (NO$_2$), 1339 (NO$_2$); ^{1}H-NMR (400 MHz, CDCl$_3$) δ: 1.85-1.90 (m, 2H, CH$_2$), 3.54 (t, J = 5.6 Hz, 4H, 2 × CH$_2$), 5.08 (br s, 1H, NH), 7.83 (d, J = 9.1 Hz, 2H, Ar), 8.20 (d, J = 9.1 Hz, 2H, Ar); ^{13}C-NMR (100 MHz, CDCl$_3$) δ: 20.4, 42.3 (2C), 123.4 (2C), 127.0 (2C), 143.2, 148.3, 152.7; *Anal.* calcd for C$_{10}$H$_{11}$N$_3$O$_2$: C, 58.53; H, 5.40; N, 20.48. Found: C, 58.61; H, 5.45; N, 20.48.

2.1.1.9 1-Methyl-2-phenyl-1,4,5,6-tetrahydropyrimidine (21)

Benzaldehyde (1.06 g, 10 mmol) and *N*-methyl- propandiamine (0.97 g, 11 mmol) was subjected to the general procedure. Product was used to next step without further purification. Yellow oil (1.49 g, 85 %); IR (neat) cm^{-1}: 1600 (C=N); ^{1}H-NMR (400 MHz, CDCl$_3$) δ: 1.92–1.98 (m, 2H, CH$_2$), 2.74 (s, 3H, NCH$_3$), 3.27 (t, J = 5.6 Hz, 2H, CH$_2$), 3.51 (t, J = 5.2 Hz, 2H, CH$_2$), 7.32–7.40 (m, 5H, Ar); ^{13}C-NMR (100 MHz, CDCl$_3$) δ: 22.0, 40.3, 45.0, 49.0, 127.9 (2C), 128.0 (2C), 128.4, 138.1, 159.1; HRMS (EI): m/z calcd for C$_{11}$H$_{13}$N$_2$ [M–H]$^{-}$ 173.1084; found: 173.1082.

2.1.1.10 General Procedure for the C–O Bond Formation (Condition B). Synthesis of 3,4-dihydro-2H-pyrimido- [1,2-c][1,3]benzoxazin-6-one (20a)

DMF (0.83 mL) and water (4.5 μL, 0.25 mmol) were added to a flask containing **18a** (40.1 mg, 0.25 mmol) and Cu(OAc)$_2$ (45.4 mg, 0.25 mmol) under an O$_2$ atmosphere. After being stirred at 130 °C for 20 min, N,N,N′,N′-tetramethyl-ethylenediamine (TMEDA, 150 μL, 1 mmol) was added. After being stirred at same temperature for 1 min, the reaction mixture was concentrated in vacuo. To a solution of residue and TMEDA (150 μL, 1 mmol) in CH$_2$Cl$_2$ (16.6 mL) was added dropwise a solution of triphosgene (77.9 mg, 0.26 mmol) in CH$_2$Cl$_2$ (1.7 mL) at 0 °C. After being stirred at rt for 1 h under an Ar atmosphere, the mixture was quenched with sat. NH$_4$Cl, and CH$_2$Cl$_2$ was removed in vacuo. The resulting mixture was made basic with 28 % NH$_4$OH. The whole was extracted with EtOAc and washed with sat. NH$_4$Cl–28 % NH$_4$OH, brine, and dried over MgSO$_4$. After concentration, the residue was purified by flash chromatography over silica gel with n-hexane–EtOAc (1:1) to give the title compound **20a** as colorless solid (35.2 mg, 70 %): mp 146–147 °C (from CHCl$_3$–n-hexane); IR (neat) cm^{-1}: 1730 (C=O), 1647 (C=N); ^{1}H-NMR (400 MHz, CDCl$_3$) δ: 1.98-2.04 (m, 2H, CH$_2$), 3.68 (t, J = 5.6 Hz, 2H, CH$_2$), 3.95 (t, J = 6.0 Hz, 2H, CH$_2$), 7.14 (d, J = 8.3 Hz, 1H, Ar), 7.23–7.30 (m, 1H, Ar), 7.48–7.51 (m, 1H, Ar), 8.02 (d, J = 7.8 Hz, 1H, Ar); ^{13}C-NMR (100 MHz, CDCl$_3$) δ: 20.3, 42.5, 44.1, 116.2, 125.0, 125.5, 127.8, 129.0, 132.9, 147.5, 150.4; HRMS (FAB): m/z calcd for C$_{11}$H$_{11}$N$_2$O$_2$ [M + H]$^+$ 203.0821; found: 203.0813.

2.1.1.11 3,4-Dihydro-2H-9-methoxypyrimido[1,2-c][1,3]benzoxazin-6-one (20b)

2-(4-Methoxyphenyl)-1,4,5,6-tetrahydropyrimidine **18b** (47.6 mg, 0.25 mmol) was subjected to the general procedure. Pale yellow solid (37.3 mg, 64 %): mp 160–161 °C (from CHCl$_3$–n-hexane); IR (neat) cm^{-1}: 1731 (C=O), 1650 (C=N); ^{1}H-NMR (500 MHz, CDCl$_3$) δ: 1.97–2.02 (m, 2H, CH$_2$), 3.64 (t, J = 5.7 Hz, 2H, CH$_2$), 3.85 (s, 3H, OCH$_3$), 3.92 (t, J = 6.0 Hz, 2H, CH$_2$), 6.59 (d, J = 2.3 Hz, 1H, Ar), 6.79 (dd, J = 8.6, 2.3 Hz, 1H, Ar), 7.90 (d, J = 8.6 Hz, 1H, Ar); ^{13}C-NMR (125 MHz, CDCl$_3$) δ: 20.5, 42.5, 44.0, 55.7, 100.0, 108.8, 112.6, 126.6, 142.7, 147.8, 151.7, 163.3; HRMS (FAB): m/z calcd for C$_{12}$H$_{13}$N$_2$O$_3$ [M + H]$^+$ 233.0926; found: 233.0921.

2.1.1.12 3,4-Dihydro-2H-9-methylpyrimido[1,2-c][1,3]benzoxazin-6-one (20c)

2-(4-Tolyl)-1,4,5,6-tetrahydropyrimidine **18c** (43.6 mg, 0.25 mmol) was subjected to the general procedure. Yellow solid (32.8 mg, 61 %): mp 153–154 °C (from

CHCl$_3$–*n*-hexane); IR (neat) cm^{-1}: 1736 (C=O), 1650 (C=N); ^{1}H-NMR (500 MHz, CDCl$_3$) δ: 1.98-2.02 (m, 2H, CH$_2$), 2.40 (s, 3H, CH$_3$), 3.66 (t, J = 5.4 Hz, 2H, CH$_2$), 3.93 (t, J = 6.0 Hz, 2H), 6.93 (s, 1H, Ar), 7.05 (d, J = 8.0 Hz, 1H, Ar), 7.88 (d, J = 8.0 Hz, 1H, Ar); ^{13}C-NMR (125 MHz, CDCl$_3$) δ: 20.4, 21.5, 42.5, 44.1, 113.4, 116.2, 125.2, 126.2, 143.0, 144.0, 148.0, 150.4; HRMS (FAB): *m/z* calcd for C$_{12}$H$_{13}$N$_2$O$_2$ [M + H]$^+$ 217.0977; found: 217.0979.

2.1.1.13 9-Bromo-3,4-dihydro-2*H*-pyrimido[1,2-*c*][1,3]benzoxazin-6-one (20d)

2-(4-Bromophenyl)-1,4,5,6-tetrahydropyrimidine **18d** (59.8 mg, 0.25 mmol) was subjected to the general procedure. Pale yellow solid (31.3 mg, 45 %): mp 206–207 °C (from CHCl$_3$–*n*-hexane); IR (neat) cm^{-1}: 1729 (C=O), 1651 (C=N); ^{1}H-NMR (500 MHz, CDCl$_3$) δ: 1.98–2.03 (m, 2H, CH$_2$), 3.65 (t, J = 5.7 Hz, 2H, CH$_2$), 3.93 (t, J = 6.0 Hz, 2H, CH$_2$), 7.31 (d, J = 1.7 Hz, 1H, Ar), 7.37 (dd, J = 8.6, 1.7 Hz, 1H, Ar), 7.87 (d, J = 8.6 Hz, 1H, Ar); ^{13}C-NMR (125 MHz, CDCl$_3$) δ: 20.3, 42.6, 44.2, 115.2, 119.4, 126.4, 126.8, 128.4, 142.1, 147.0, 150.7; HRMS (FAB): *m/z* calcd for C$_{11}$H$_{10}$BrN$_2$O$_2$ [M + H]$^+$ 280.9926; found: 280.9922.

2.1.1.14 3,4-Dihydro-2*H*-9-(methoxycarbonyl)pyrimido[1,2-*c*][1,3]benzoxazin-6-one (20e)

2-[(4-Methoxycarbonyl)phenyl]-1,4,5,6-tetrahydropyrimidine **18e** (54.6 mg, 0.25 mmol) was subjected to the general procedure. Pale yellow solid (30.2 mg, 46 %): mp 136–137 °C (from CHCl$_3$–*n*-hexane); IR (neat) cm^{-1}: 1741 (C=O), 1718 (C=O), 1644 (C=N); ^{1}H-NMR (500 MHz, CDCl$_3$) δ: 2.00–2.05 (m, 2H, CH$_2$), 3.70 (t, J = 5.4 Hz, 2H, CH$_2$), 3.94–3.96 (m, 5H, CH$_2$, OMe), 7.78 (d, J = 1.4 Hz, 1H, Ar), 7.88 (dd, J = 8.6, 1.4 Hz, 1H, Ar), 8.09 (d, J = 8.6 Hz, 1H, Ar); ^{13}C-NMR (125 MHz, CDCl$_3$) δ: 20.2, 42.5, 44.4, 52.6, 117.6, 119.7, 125.6, 125.8, 134.3, 142.2, 147.1, 150.2, 165.4; HRMS (FAB): *m/z* calcd for C$_{13}$H$_{13}$N$_2$O$_4$ [M + H]$^+$ 261.0875; found: 261.0874.

2.1.1.15 3,4-Dihydro-2*H*-9-(trifluoromethyl)pyrimido[1,2-*c*][1,3]benzoxazin-6-one (20f)

2-[4-(Trifluoromethyl)phenyl]-1,4,5,6-tetrahydropyrimidine **18f** (57.1 mg, 0.25 mmol) was subjected to the general procedure. Yellow solid (28.8 mg, 43 %): mp 141–142 °C (from CHCl$_3$–*n*-hexane); IR (neat) cm^{-1}: 1739 (C=O), 1650 (C=N); ^{1}H-NMR (500 MHz, CDCl$_3$) δ: 2.00–2.05 (m, 2H, CH$_3$), 3.70 (t, J = 5.7 Hz, 2H, CH$_2$), 3.95 (t, J = 6.0 Hz, 2H, CH$_2$), 7.40 (d, J = 1.1 Hz, 1H, Ar), 7.49 (dd, J = 8.0, 1.1 Hz, 1H, Ar), 8.15 (d, J = 8.0 Hz, 1H, Ar); ^{13}C-NMR

(100 MHz, CDCl$_3$) δ: 20.3, 42.7, 44.5, 113.9 (q, $J = 4.1$ Hz), 119.3, 121.6 (q, $J = 3.6$ Hz), 124.5, 126.7, 134.7 (q, $J = 33.7$ Hz), 141.8, 146.9, 150.4; ^{19}F-NMR (500 MHz, CDCl$_3$) δ: –63.0; HRMS (FAB): m/z calcd for $C_{12}H_{10}F_3N_2O_2$ [M + H]$^+$ 271.0694; found: 271.0692.

2.1.1.16 3,4-Dihydro-2H-9-nitropyrimido[1,2-c][1,3]benzoxazin-6-one (20g)

2-(4-Nitrophenyl)-1,4,5,6-tetrahydropyrimidine **18g** (51.3 mg, 0.25 mmol) was subjected to the general procedure. Yellow solid (11.9 mg, 19 %): mp 235–236 °C (from CHCl$_3$–n-hexane); IR (neat) cm^{-1}: 1732 (C=O), 1641 (C=N), 1531 (NO$_2$), 1349 (NO$_2$). ^{1}H-NMR (400 MHz, CDCl$_3$) δ: 2.01–2.07 (m, 2H, CH$_2$), 3.72 (t, $J = 5.6$ Hz, 2H, CH$_2$), 3.96 (t, $J = 6.0$ Hz, 2H, CH$_2$), 8.00 (d, $J = 2.2$ Hz, 2H, Ar), 8.08 (dd, $J = 8.8$, 2.2 Hz, 1H, Ar), 8.22 (d, $J = 8.8$ Hz, 1H, Ar); ^{13}C-NMR (100 MHz, CDCl$_3$) δ: 20.1, 42.6, 44.5, 112.2, 119.4, 121.4, 127.1, 141.3, 146.4, 150.3, 150.5; HRMS (FAB): m/z calcd for $C_{11}H_{10}N_3O_4$ [M + H]$^+$ 248.0671; found: 248.0670.

2.1.1.17 C–N Bond Formation with BocNH$_2$. Synthesis of 3,4-Dihydro-2H,6H-pyrimido[1,2-c]quinazolin-6(7H)-one (23a)

DMF (0.83 mL) was added to a flask containing **18a** (40.1 mg, 0.25 mmol), Cu(OAc)$_2$ (45.4 mg, 0.25 mmol) and *tert*-butyl carbamate (87.9 mg, 0.75 mmol) under an O$_2$ atmosphere. After being stirred at 100 °C for 40 min, the mixture was concentrated *in vacuo*. The residue was purified by flash chromatography over aluminum oxide with CHCl$_3$–MeOH (1:0 to 99:1) to give **23a** as colorless solid (26.5 mg, 53 %): mp 250–251 °C (from CHCl$_3$–n-hexane); IR (neat) cm^{-1}: 1682 (C=O), 1616 (C=N); ^{1}H-NMR (400 MHz, CDCl$_3$) δ: 1.95–2.00 (m, 2H, CH$_2$), 3.67 (t, $J = 5.6$ Hz, 2H, CH$_2$), 3.94 (t, $J = 6.0$ Hz, 2H, CH$_2$), 6.86 (d, $J = 8.0$ Hz, 1H, Ar), 7.09–7.13 (m, 1H, Ar), 7.38–7.42 (m, 1H, Ar), 8.07 (d, $J = 8.0$ Hz, 1H, Ar), 8.30 (br s, 1H, NH); ^{13}C-NMR (100 MHz, CDCl$_3$) δ: 20.3, 40.8, 44.5, 114.6, 116.5, 123.0, 125.8, 132.0, 136.5, 145.7, 151.2; HRMS (FAB): m/z calcd for $C_{11}H_{12}N_3O$ [M + H]$^+$ 202.0980; found: 202.0988.

2.1.1.18 C–N Bond Formation with TsNH$_2$. Synthesis of 7-Tosyl-3,4-dihydro-2H,6H-pyrimido[1,2-c]quinazolin-6(7H)-one (23b)

DMF (0.83 mL) was added to a flask containing **18a** (40.1 mg, 0.25 mmol), Cu(OAc)$_2$ (45.4 mg, 0.25 mmol) and p-toluene sulfonamide (85.6 mg, 0.5 mmol) under an O$_2$ atmosphere. After being stirred at 130 °C for 20 min, the mixture was concentrated *in vacuo*. The residue was subjected to flash chromatography over aluminum oxide with CHCl$_3$–MeOH (95:5) to give crude *ortho*-amidated compound. To a solution of the *ortho*-amidated compound and Et$_3$N (145 μL, 1.0 mmol) in CH$_2$Cl$_2$ (16.6 mL) was added dropwise a solution of triphosgene (77.9 mg, 0.26 mmol) in CH$_2$Cl$_2$ (1.7 mL) at 0 °C. After being stirred at rt for 1 h under an Ar atmosphere, the mixture was quenched with sat. NaHCO$_3$, and CH$_2$Cl$_2$ was removed *in vacuo*. The whole was extracted with EtOAc and washed with sat. NaHCO$_3$, brine, and dried over MgSO$_4$. After concentration, the residue was purified by flash chromatography over silica gel with *n*-hexane–EtOAc (1:1) to give **23b** as colorless solid (42.2 mg, 47 %): mp 159–161 °C (from CHCl$_3$–*n*-hexane); IR (neat) cm^{-1}: 1695 (C=O), 1644 (C=N); ^{1}H-NMR (400 MHz, CDCl$_3$) δ: 1.85–1.91 (m, 2H, CH$_2$), 2.46 (s, 3H, CH$_3$), 3.63 (t, *J* = 5.7 Hz, 2H, CH$_2$), 3.75 (t, *J* = 6.2 Hz, 2H, CH$_2$), 7.27–7.31 (m, 1H, Ar), 7.37 (d, *J* = 8.3 Hz, 2H, Ar), 7.48–7.53 (m, 1H, Ar), 7.87 (d, *J* = 8.5 Hz, 1H, Ar), 8.03 (d, *J* = 8.3 Hz, 2H, Ar), 8.07 (dd, *J* = 8.0, 1.7 Hz, 1H, Ar); ^{13}C-NMR (100 MHz, CDCl$_3$) δ: 20.5, 21.8, 41.8, 44.6, 120.3, 121.0, 125.7, 126.4, 128.4 (2C), 129.8 (2C), 131.3, 134.6, 136.7, 144.5, 145.4, 148.3; HRMS (FAB): *m/z* calcd for C$_{18}$H$_{18}$N$_3$O$_3$S [M + H]$^+$ 356.1069; found: 356.1074.

2.1.1.19 C–S Bond Formation with CS$_2$. Synthesis of 3,4-Dihydro-2*H*,6*H*-pyrimido[1,2-*c*][1,3]benzothiazine-6-thione (24)

To a solution of **18a** (40.1 mg, 0.25 mmol), Cu(OAc)$_2$ (45.4 mg, 0.25 mmol) in 1,4-dioxane (0.83 mL) was added CS$_2$ (0.045 mL, 0.75 mmol) under an O$_2$ atmosphere. After being stirred at 130 °C for 15 min, the mixture was concentrated. The residue was purified by flash chromatography over silica gel with *n*-hexane–EtOAc (9:1) to give the title compound **24** as pale yellow solid (6.6 mg, 11 %): mp 139–141 °C (from CHCl$_3$–*n*-hexane); IR (neat) cm^{-1}: 1624 (C=N); ^{1}H-NMR (400 MHz, CDCl$_3$) δ: 2.01–2.07 (m, 2H, CH$_2$), 3.76 (t, *J* = 5.6 Hz, 2H, CH$_2$), 4.45 (t, *J* = 6.2 Hz, 2H, CH$_2$), 7.03 (dd, *J* = 7.8, 1.5 Hz, 1H, Ar), 7.28–7.33 (m, 1H, Ar), 7.41 (ddd, *J* = 8.0, 7.6, 1.5 Hz, 1H, Ar), 8.20 (dd, *J* = 8.0, 1.2 Hz, 1H, Ar); ^{13}C-NMR (125 MHz, CDCl$_3$) δ: 21.6, 45.5, 48.6, 121.6, 126.5, 127.5, 128.9, 131.1, 131.8, 144.2, 189.8; *Anal.* calcd for C$_{11}$H$_{10}$N$_2$S$_2$: C, 56.38; H, 4.30; N, 11.95. Found: C, 56.23; H, 4.44; N, 11.85.

2.2 S$_N$Ar-Type C–S, C–N, or C–O Bond Formation with Heterocumulenes

The C–H functionalization methodology described in Chap 2.1 provides a facile access to tricyclic heterocycles related to PD 404182; however, C–S bond formation failed to synthesize several derivatives with the pyrimido[1,2-*c*][1,3]benzothiazin-6-imine scaffold because of the low yield.

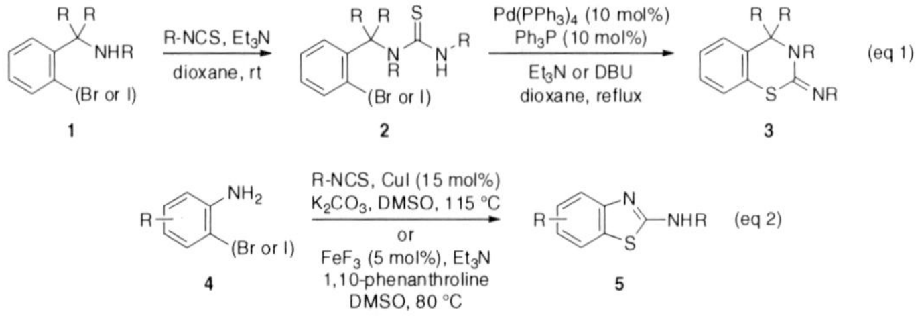

Scheme 2.10 Examples of transition-metal-catalyzed coupling of haloarene and heterocumulene

The transition-metal-catalyzed carbon–heteroatom bond formations such as Ullmann–Goldberg reactions and Buchwald–Hartwig cross coupling are becoming a powerful methods for construction of various heterocycles.[11,12] Orain and co-workers have reported a Pd-catalyzed intramolecular S-arylation of thioureas **2** to yield 3,4-dihydro-2H-benzo[e][1,3]thiazin-2-imine derivatives **3** (Scheme 2.10, eq 1) [43] Thioureas were easily obtained by the reaction of 2-halobenzyl-amine derivatives **1** and isothiocyanates. Bao and co-workers have reported the formation of 2-amino-benzothiazoles **5** by CuI-catalyzed coupling of 2-haloanilines **4** and isothiocyanates (eq 2) [44] Li and co-workers have revealed that thses 2-amino-benzothiazole formation reactions were assisted by an Fe(III) catalyst [45]. These reactions proceed via nucleophilic addition of aniline to isothiocyanate followed by transition-metal-catalyzed intramolecular S-arylation.

With these findings [38–53], the author investigated the transition-metal (Pd, Cu, Fe, etc.) catalyzed coupling of haloarenes **6aa** and heterocumulenes. During examination of the coupling reaction of **6aa** with CS_2 (Scheme 2.11, eq 1), the author noticed that the desired compound **7a** was formed without using a transition-metal catalyst (eq 2).

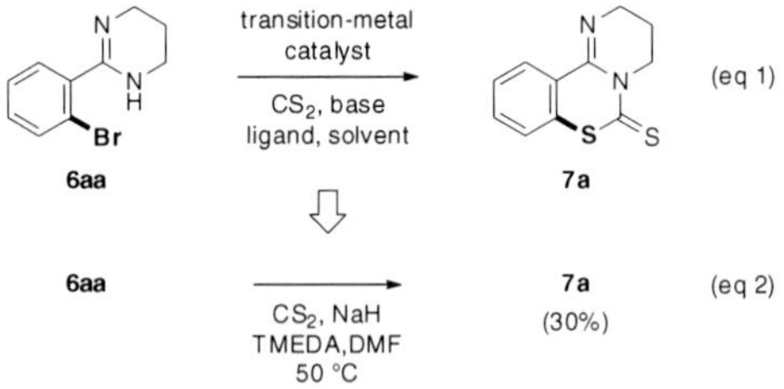

Scheme 2.11 Synthesis of PD 404182 Derivatives **7a** by the coupling of haloarene and heterocumulene. TMEDA N,N,N′,N′-tetramethylethylenediamine

---

[11] For reviews on transition-metal-catalyzed carbon–heteroatom bond formation, see [38– 41].

[12] For recent examples of the transition-metal-catalyzed coupling reaction of haloarene and hetelocumulenes, see [41–53].

There are several reports of transition-metal-free C–S bond formation. Kobayashi and co-workers have reported the coupling reaction of 2-chloropyridine derivatives **8** with CS_2 (Scheme 2.12, eq 1) [57]. In this reaction, *S*-functionality is introduced at electronically activated C-2 position through the aromatic nucleophilic substitution (S_NAr) reaction. This report encouraged the author to examine the coupling of halo-arenes **6aa** and heterocumulenes by S_NAr reaction for the synthesis of PD 404182 derivatives [54–61].[13] After the authors' report, [59] Xi and co-workers reported [59] DBU-promoted tandem reaction of 2-haloanilines **10** and CS_2 (eq 2) [60].

The author initially examined the reaction of **6aa** [6] with 5 equiv of sodium hydride and CS_2 (Table 2.3). The desired reaction efficiently proceeded in DMF to give **7a** in 75 % yield (entry 1). In contrast, when acetonitrile or THF was used as

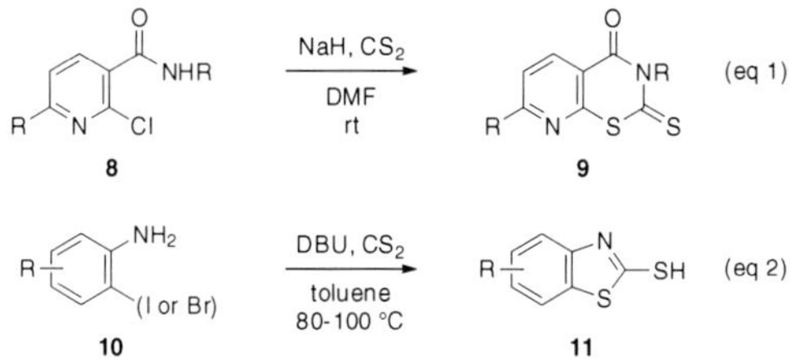

Scheme 2.12 Examples of transition-metal-free coupling of haloarene and heterocumulene

Table 2.3 Optimization of reaction conditions with CS_2[a]

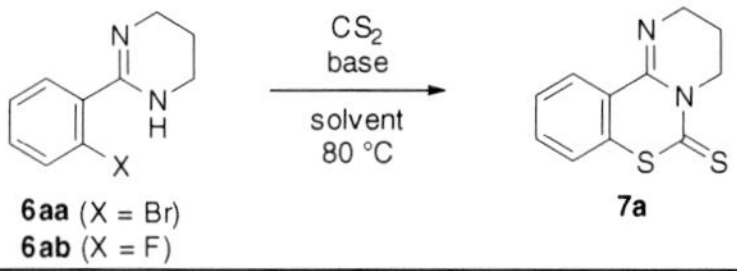

Entry	X	Base (equiv)	Solvent	Time (h)	Yield (%)[b]
1	Br	NaH (5)	DMF	6	75
2	Br	NaH (5)	MeCN	4	Trace
3	Br	NaH (5)	THF	4	Trace
4	Br	NaH (2)	DMF	12	88
5	Br	None	DMF	12	12
6	Br	Et_3N (2)	DMF	12	Trace
7	Br	KH (2)	DMF	6	Trace
8	Br	NaOt-Bu (2)	DMF	6	27
9	F	NaH (2)	DMF	12	86

[a] All reactions were carried out at 80 °C with 2 or 5 equiv of CS_2 (corresponding to the base loading)
[b] Isolated yields

---

[13] For examples of the transition-metal-free coupling reaction of haloarene and hetelocumulenes, see [58–61].

the solvent instead of DMF, yields of **7a** decreased considerably (entries 2 and 3). A decreasing amount of sodium hydride and CS$_2$ (2.0 equiv) slightly improved the yield of **7a** (88 %) under the reaction for 12 h (entry 4). The reaction in the absence of sodium hydride provided a yield of **7a** of only 12 % (entry 5). The author next screened several bases such as triethylamine, potassium hydride[14] and sodium *tert*-butoxide (entries 6-8): sodium hydride was the most effective (entry 4). The fluoride **6ab** gave a comparable result with the bromide **6aa** to afford **7a** in 86 % yield under optimized conditions (entry 9).

With knowledge of the optimized conditions, the author examined the reaction of several substituted substrates (Table 2.4). Substrates **6b–d** having a methoxy, methyl, or fluoro group at the 4-position provided the corresponding cyclized

Table 2.4 Reaction of substituted 2-(2-halophenyl)-1,4,5,6-tetrahydropyrimidines[a]

Entry	Substrate	Product	Yield (%)[b]
	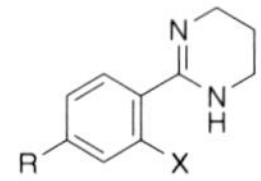	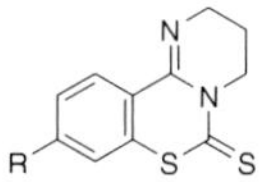	
1	**6b** (R = OMe, X = F)	**7b** (R = OMe)	95
2	**6c** (R = Me, X = Br)	**7c** (R = Me)	88
3	**6d** (R = F, X = Br)	**7d** (R = F)	76
4	**6e** (R = NO$_2$, X = F)	**7e** (R = NO$_2$)	(73)[c]
	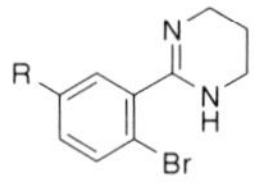	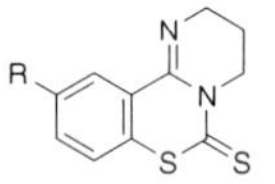	
5	**6f** (R = OMe)	**7f** (R = OMe)	17
6	**6g** (R = NO$_2$)	**7g** (R = NO$_2$)	(57)[c]
7	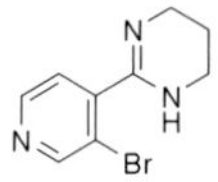 **12**	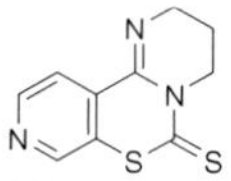 **13**	18
8	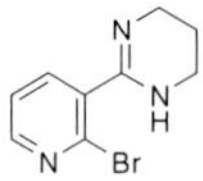 **14**	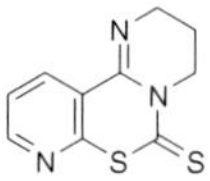 **15**	71
9	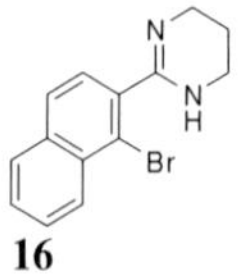 **16**	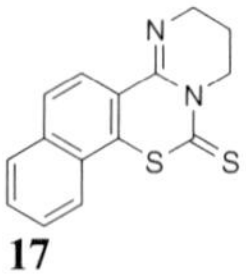 **17**	>99

[a] Unless otherwise stated, reactions were carried out with CS$_2$ (2.0 equiv) and NaH (2.0 equiv) in DMF at 80 °C for 12 h
[b] Isolated yields
[c] Yields in parentheses indicate those of the reactions at rt

[14] A reason for the significant countercation effect (NaH vs. KH) on the reactivity is unclear.

Table 2.5 Reaction with isothiocyanates or isocyanates[a]

Entry	Substrate	R-NCX	Product	Yield (%)[b]
1	**6aa** (X = Br)	BnNCS	**18**	82
2	**6ab** (X = F)			97
3[c]	**6ab**	t-BuNCS	**19**	62[d,e]
4	**6ab**	BnNCO	**20**	>99
5[c]	**6ab**	t-BuNCO	**21** (Y = Nt-Bu, Z = O) **22** (Y = O, Z = Nt-Bu)	54 18[e]
6	**6ab**	PhNCO	**23**	>99
7	**24**	t-BuNCS	**25**	49[e]

[a] Unless otherwise stated, reactions were carried out with R-NCX (2.0 equiv) and NaH (2.0 equiv) in DMF at rt for 2–3 h
[b] Isolated yields
[c] These reactions were carried out at 80 °C
[d] A trace amount of regioisomeric N-arylation product was also formed
[e] Isolated as a single isomer

products **7b–d** in good-to-excellent yields (76–95 %, entries 1–3). Whereas the reaction of **6e** bearing the 4-nitro group at 80 °C resulted in the formation of a complex mixture, the reaction at room temperature gave the cyclization product **7e** in 73 % yield (entry 4). A methoxy group on the 5-position considerably diminished the reactivity, affording **7f** in only 17 % yield (entry 5). This was presumably due to increased electron density at the carbon substituted by a bromine atom. In the case of **6g** bearing a 5-nitro group, the corresponding product **7g** was obtained by the reaction at room temperature (entry 6), similarly to **6e** (entry 4). Pyridine derivatives **12** and **14** showed different reactivity depending on the position of the nitrogen atom: the 2-bromopyridine derivative **14** gave a better result (71 %, entry 8) than the 3-bromopyridine derivative **12** (18 %, entry 7). The naphthalene derivative **16** afforded the tetracyclic compound **17** in quantitative yield (entry 9).

To further expand this methodology for the construction of other heterocyclic frameworks, the author investigated the reaction using isothiocyanates or isocyanates[15] as heterocumulene (Table 2.5). When benzylisothiocyanate was employed, the reaction of **6aa** or **6ab** efficiently proceeded to give the corresponding *N*-arylated product **18** in 82 % or 97 % yields, respectively (entries 1 and 2). The reaction with *tert*-butylisothiocyanate exclusively furnished an *S*-arylated product **19** as a single isomer (entry 3). These results indicate that the regioselectivity of the reaction can be perfectly switched by changing a substituent on the nitrogen atom. As expected, the reaction of **6ab** with benzylisocyanate provided an *N*-arylated product **20** in quantitative yield (entry 4) as in the case with isothiocyanate (entries 1 and 2). Interestingly, *tert*-butylisocyanate showed moderate selectivity to mainly afford an *N*-arylation product **21** (54 %), formed by the arylation at the more bulky position, as well as an *O*-arylation product **22** (18 %, entry 5). Phenylisocyanate also provided an *N*-arylated product **23** (entry 6). The 2-phenylimidazoline derivative **24** (a 5-membered-ring amidine congener) also provided the corresponding *S*-arylated product **25** in a slightly decreased yield (49 %, entry 7).

This reaction would proceed via a nucleophilic addition of the amidine moiety to heterocumulene followed by an intramolecular S_NAr reaction[16,17] of the resulting adducts such as **B** (Scheme 2.13). Nonactivated aromatic rings efficiently reacted under relatively mild conditions, so two molecules of the heterocumulene may be involved in the reaction to form the intermediate **C** in which the amidine moiety can be a more powerful electron-withdrawing group suitable for the S_NAr-type reaction. The regioselectivity in the nucleophilic attack on the aromatic ring (Y vs. Z) is controlled by a subtle balance of inherent nucleophilicity and steric hindrance of these functionalities.

The author finally focused on the synthesis of PD 404182 (**26**) (Scheme 2.14). Hydrolysis of the carbamodithioate derivative **7a** followed by treatment with cyanogen bromide [3] readily afforded the desired compound **26**. The same compound was also obtained in a single step by heating compound **19** in trifluoroacetic acid in the presence of molecular sieves.

In conclusion, the author developed a simple and practical synthetic method for tricyclic heteroarenes related to PD 404182. This reaction provides divergent access to several related heterocycles under mild conditions without a powerful activating group.

---

[15] For related reactions of electron-deficient (haloaryl)isothiocyanates, see [62–64].

[16] For reviews on nucleophilic aromatic substitution reaction, see [65, 66].

[17] For examples on nucleophilic aromatic substitution reaction, see [67–73].

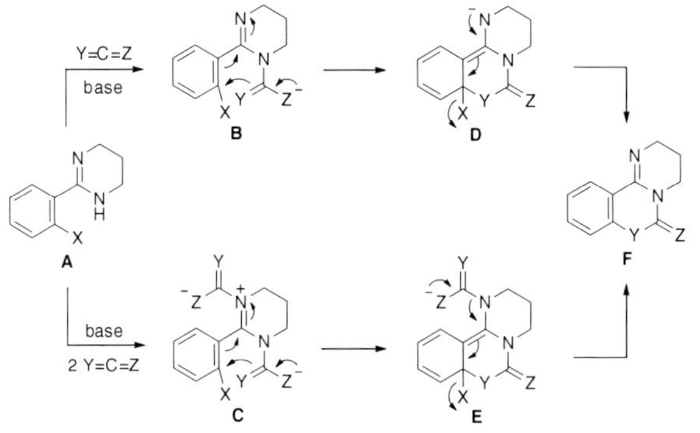

Scheme 2.13 Proposed reaction mechanisms

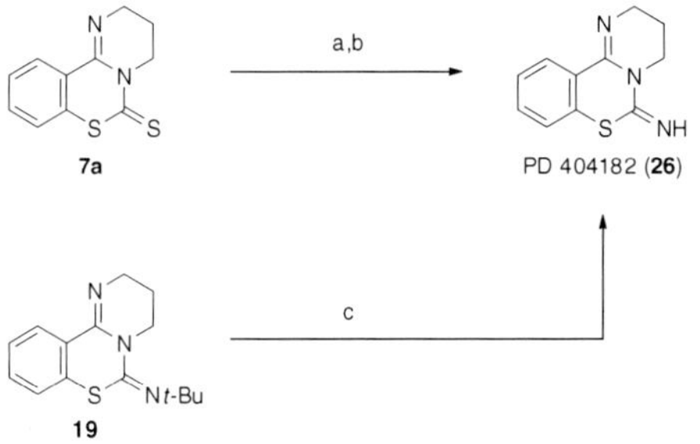

Scheme 2.14 Synthesis of PD 404182. Reagents and conditions: (**a**) NaOH, MeOH-H$_2$O (9:1), reflux; (**b**) BrCN, EtOH, reflux, 61 % (2 steps); (**c**) TFA, MS4Å, CHCl$_3$, reflux, 85 %

2.2.1 Experimental Section

2.2.1.1 General Methods

All moisture-sensitive reactions were performed using syringe-septum cap techniques under an Ar atmosphere and all glasswares were dried in an oven at 80 °C for 2 h prior to use. Melting points were measured by a hot stage melting point apparatus (uncorrected). For flash chromatography, Wakogel C-300E (Wako) or aluminum oxide 90 standardized (Merck) was employed. ^{1}H-NMR spectra were recorded using a JEOL AL-400 or a JEOL ECA-500 spectrometer, and chemical shifts are reported in δ (ppm) relative to Me$_4$Si (CDCl$_3$) as internal standards. ^{13}C-NMR spectra were recorded using a JEOL AL-400 or JEOL ECA-500 spectrometer and referenced to the residual CHCl$_3$ signal. ^{19}F–NMR spectra were recorded using a JEOL ECA-500 and referenced to the internal CFCl$_3$ (δ $_F$ 0.00 ppm). ^{1}H-NMR spectra are tabulated as follows: chemical shift, multiplicity (b = broad, s = singlet, d = doublet, t = triplet, q = quartet, m = multiplet), coupling constant(s), and number of protons. Exact mass (HRMS) spectra were recorded on a JMS-HX/HX 110A mass spectrometer. Infrared (IR) spectra were obtained on a JASCO FT/IR-4100 FT-IR spectrometer with JASCO ATR PRO410-S.

2.2.1.2 General Procedure for Preparation of the Substrates.
2-(2-Bromophenyl)-1,4,5,6-tetrahydropyrimidine (6aa)

To a solution of 2-bromobenzaldehyde (5.55 g, 30.0 mmol) in t-BuOH (280 mL) was added propylenediamine (2.45 g, 33.0 mmol). The mixture was stirred at 70 °C for 30 min, and then K_2CO_3 (12.4 g, 90.0 mmol) and I_2 (9.52 g, 37.5 mmol) were added. After being stirred at same temperature for 3 h, the mixture was quenched with sat. Na_2SO_3 until the iodine color disappeared. The organic layer was separated and concentrated *in vacuo*. The resulting solid was dissolved with H_2O, and then pH was adjusted to 12–14 with 2 N NaOH. The whole was extracted with $CHCl_3$. The extract was dried over Na_2SO_4. After concentration, the resulting solid was recrystallized from $CHCl_3$–n-hexane to give the compound **6aa** as colorless crystals (6.63 g, 92 %): mp 136–137 °C; IR (neat) cm^{-1}: 1625 (C=N); ^{1}H-NMR (500 MHz, $CDCl_3$) δ: 1.81-1.86 (m, 2H, CH_2), 3.42 (t, J = 5.7 Hz, 4H, 2 × CH_2), 4.83 (br s, 1H, NH), 7.18 (ddd, J = 8.0, 7.7, 1.7 Hz, 1H, Ar), 7.27-7.31 (m, 1H, Ar), 7.41 (dd, J = 7.7, 1.7 Hz, 1H, Ar), 7.53 (d, J = 8.0 Hz, 1H, Ar); ^{13}C-NMR (125 MHz, $CDCl_3$) δ: 20.5, 42.2 (2C), 120.7, 127.3, 129.9, 130.2, 132.7, 139.3, 155.3; HRMS (FAB): m/z calcd for $C_{10}H_{12}BrN_2$ $[M + H]^+$ 239.0184; found: 239.0185.

2.2.1.3 2-(2-Fluorophenyl)-1,4,5,6-tetrahydropyrimidine (6ab)

2-Fluorobenzaldehyde (1.24 g, 10.0 mmol) was subjected to the general procedure. Colorless crystals (1.28 g, 71 %): mp 112–113 °C (from $CHCl_3$–n-hexane); IR (neat) cm^{-1}: 1629 (C=N); ^{1}H-NMR (400 MHz, $CDCl_3$) δ: 1.82–1.88 (m, 2H, CH_2), 3.49 (t, J = 5.7 Hz, 4H, 2 × CH_2), 5.33 (br s, 1H, NH), 7.03 (ddd, J = 11.9, 8.2, 1.1 Hz, 1H, Ar), 7.14 (ddd, J = 7.8, 7.8, 1.1 Hz, 1H, Ar), 7.29–7.34 (m, 1H, Ar), 7.80 (ddd, J = 7.8, 7.8, 2.0 Hz, 1H, Ar); ^{13}C-NMR (100 MHz, $CDCl_3$) δ: 20.6, 42.2 (2C), 115.8 (d, J = 23.2 Hz), 124.2 (d, J = 3.3 Hz), 124.5 (d, J = 11.6 Hz), 130.5 (d, J = 3.3 Hz), 130.7 (d, J = 8.3 Hz), 151.5, 160.1 (d, J = 247.5 Hz); ^{19}F-NMR (500 MHz, $CDCl_3$) δ: –116.1; *Anal.* calcd for $C_{10}H_{11}FN_2$: C, 67.40; H, 6.22; N, 15.72. Found: C, 67.15; H, 6.32; N, 15.63.

2.2.1.4 2-(2-Fluoro-4-methoxyphenyl)-1,4,5,6-tetrahydropyrimidine (6b)

2-Fluoro-4-methoxybenzaldehyde (0.77 g, 5.0 mmol) was subjected to the general procedure. Pale yellow crystals (0.70 g, 67 %): mp 77 °C (from CHCl$_3$–n-hexane); IR (neat) cm^{-1}: 1623 (C=N); ^{1}H-NMR (400 MHz, CDCl$_3$) δ: 1.82-1.88 (m, 2H, CH$_2$), 3.48 (t, J = 5.7 Hz, 4H, 2 × CH$_2$), 3.79 (s, 3H, OCH$_3$), 5.13 (br s, 1H, NH), 6.56 (dd, J = 13.8, 2.6 Hz, 1H, Ar), 6.69 (dd, J = 8.8, 2.6 Hz, 1H, Ar), 7.77 (dd, J = 8.8, 8.8 Hz, 1H, Ar); ^{13}C-NMR (125 MHz, CDCl$_3$) δ: 20.8, 42.2 (2C), 55.5, 101.5 (d, J = 27.6 Hz), 110.2 (d, J = 2.4 Hz), 116.7 (d, J = 10.8 Hz), 131.3 (d, J = 6.0 Hz), 151.4 (d, J = 2.4 Hz), 160.7 (d, J = 205.1 Hz), 161.7 (d, J = 30.0 Hz); ^{19}F-NMR (500 MHz, CDCl$_3$) δ: –113.8; *Anal.* calcd for C$_{11}$H$_{13}$FN$_2$O: C, 63.45; H, 6.29; N, 13.45. Found: C, 63.38; H, 6.29; N, 13.49.

2.2.1.5 2-(2-Bromo-4-methylphenyl)-1,4,5,6-tetrahydropyrimidine (6c)

2-Bromo-4-methylbenzaldehyde (1.00 g, 5.0 mmol) was subjected to the general procedure. Colorless crystals (1.11 g, 88 %): mp 128–129 °C (from CHCl$_3$–n-hexane); IR (neat) cm^{-1}: 1639 (C=N); ^{1}H-NMR (400 MHz, CDCl$_3$) δ: 1.83-1.89 (m, 2H, CH$_2$), 2.31 (s, 3H, CH$_3$), 3.46 (t, J = 5.7 Hz, 4H, 2 × CH$_2$), 4.65 (br s, 1H, NH), 7.07–7.10 (m, 1H, Ar), 7.27–7.35 (m, 2H, Ar); ^{13}C-NMR (100 MHz, CDCl$_3$) δ: 20.7, 20.8, 42.3 (2C), 120.4, 128.1, 130.1, 133.2, 136.5, 140.3, 155.3; *Anal.* calcd for C$_{11}$H$_{13}$BrN$_2$: C, 52.19; H, 5.18; N, 11.07. Found: C, 52.37; H, 5.21; N, 11.12.

2.2.1.6 2-(2-Bromo-4-fluorophenyl)-1,4,5,6-tetrahydropyrimidine (6d)

2-Bromo-4-fluorobenzaldehyde (1.02 g, 5.0 mmol) was subjected to the general procedure. Colorless crystals (1.25 g, 97 %): mp 130 °C (from CHCl$_3$–n-hexane); IR (neat) cm^{-1}: 1623 (C=N); ^{1}H-NMR (500 MHz, CDCl$_3$) δ: 1.83-1.87 (m, 2H, CH$_2$), 3.44 (t, J = 6.0 Hz, 4H, 2 × CH$_2$), 4.42 (br s, 1H, NH), 7.01 (ddd, J = 8.6, 8.3, 2.7 Hz, 1H, Ar), 7.27 (dd, J = 8.3, 2.7 Hz, 1H, Ar), 7.40 (dd, J = 8.6, 5.7 Hz, 1H, Ar); ^{13}C-NMR (125 MHz, CDCl$_3$) δ: 20.5, 42.3 (2C), 114.6 (d, J = 20.4 Hz), 120.0 (d, J = 24.0 Hz), 121.1 (d, J = 9.6 Hz), 131.4 (d, J = 8.4 Hz), 135.6 (d, J = 3.6 Hz), 154.5, 162.2 (d, J = 251.9 Hz); ^{19}F-NMR (500 MHz, CDCl$_3$) δ: –116.1. *Anal.* calcd for C$_{10}$H$_{10}$BrFN$_2$: C, 46.72; H, 3.92; N, 10.90. Found: C, 46.64; H, 3.87; N, 10.97.

2.2.1.7 2-(2-Fluoro-4-nitrophenyl)-1,4,5,6-tetrahydropyrimidine (6e)

2-Fluoro-4-nitrobenzaldehyde (0.68 g, 4.0 mmol) was subjected to the general procedure. Yellow crystals (0.69 g, 77 %): mp 141–142 °C (from CHCl$_3$–n-

hexane); IR (neat) cm^{-1}: 1625 (C=N), 1603 (NO$_2$), 1519 (NO$_2$); ^{1}H-NMR (400 MHz, CDCl$_3$) δ: 1.84-1.90 (m, 2H, CH$_2$), 3.50 (t, J = 5.7 Hz, 4H, 2 × CH$_2$), 5.51 (br s, 1H, NH), 7.90–8.02 (m, 3H, Ar); ^{13}C-NMR (125 MHz, CDCl$_3$) δ: 20.4, 44.8 (2C), 112.0 (d, J = 30.0 Hz), 119.2 (d, J = 3.6 Hz), 130.5 (d, J = 12.0 Hz), 131.9 (d, J = 3.6 Hz), 148.8 (d, J = 9.6 Hz), 149.8, 159.4 (d, J = 251.9 Hz); ^{19}F-NMR (500 MHz, CDCl$_3$) δ: –111.8; *Anal.* calcd for C$_{10}$H$_{10}$FN$_3$O$_2$: C, 53.81; H, 4.52; N, 18.83. Found: C, 54.05; H, 4.53; N, 19.05.

2.2.1.8 2-(2-Bromo-5-methoxyphenyl)-1,4,5,6-tetrahydropyrimidine (6f)

2-Bromo-5-methoxybenzaldehyde (0.86 g, 4.0 mmol) was subjected to the general procedure. Colorless crystals (0.98 g, 91 %): mp 124 °C (from CHCl$_3$–n-hexane); IR (neat) cm^{-1}: 1626 (C=N); ^{1}H-NMR (500 MHz, CDCl$_3$) δ: 1.84–1.89 (m, 2H, CH$_2$), 3.46 (t, J = 6.0 Hz, 4H, 2 × CH$_2$), 3.79 (s, 3H, OCH$_3$), 4.63 (br s, 1H, NH), 6.76 (dd, J = 9.2, 3.2 Hz, 1H, Ar), 6.98 (d, J = 3.2 Hz, 1H, Ar), 7.39 (d, J = 9.2 Hz, 1H, Ar); ^{13}C-NMR (125 MHz, CDCl$_3$) δ: 20.6, 42.3 (2C), 55.5, 110.9, 115.0, 116.9, 133.5, 140.0, 155.3, 158.9; *Anal.* calcd for C$_{11}$H$_{13}$BrN$_2$O: C, 49.09; H, 4.87; N, 10.41. Found: C, 49.21; H, 4.84; N, 10.44.

2.2.1.9 2-(2-Bromo-5-nitrophenyl)-1,4,5,6-tetrahydropyrimidine (6g)

2-Bromo-5-nitrobenzaldehyde (0.58 g, 2.5 mmol) was subjected to the general procedure. Yellow crystals (0.41 g, 58 %): mp 139–141 °C (from CHCl$_3$–n-hexane); IR (neat) cm^{-1}: 1631 (C=N), 1608 (NO$_2$), 1524 (NO$_2$); ^{1}H-NMR (400 MHz, CDCl$_3$) δ: 1.87–1.93 (m, 2H, CH$_2$), 3.50 (t, J = 5.7 Hz, 4H, 2 × CH$_2$), 4.59 (br s, 1H, NH), 7.72 (d, J = 8.8 Hz, 1H, Ar), 8.03 (dd, J = 8.8, 2.7 Hz, 1H, Ar), 8.27 (d, J = 2.7 Hz, 1H, Ar); ^{13}C-NMR (100 MHz, CDCl$_3$) δ: 20.5, 42.4 (2C), 124.4, 125.3, 128.3, 134.0, 140.5, 147.1, 153.4; *Anal.* calcd for C$_{10}$H$_{10}$BrN$_3$O$_2$: C, 42.27; H, 3.55; N, 14.79. Found: C, 42.55; H, 3.80; N, 14.52.

2.2.1.10 2-(3-Bromopyridin-4-yl)-1,4,5,6-tetrahydropyrimidine (12)

3-Bromoisonicotinaldehyde (0.93 g, 5.0 mmol) was subjected to the general procedure. Yellow solid (0.73 g, 61 %): mp 141 °C (from CHCl$_3$–n-hexane); IR (neat) cm^{-1}: 1630 (C=N); ^{1}H-NMR (400 MHz, CDCl$_3$) δ: 1.84–1.90 (m, 2H, CH$_2$), 3.46 (t, J = 5.7 Hz, 4H, 2 × CH$_2$), 4.93 (br s, 1H, NH), 7.35 (d, J = 4.6 Hz, 1H, Ar), 8.50 (d, J = 4.6 Hz, 1H, Ar), 8.70 (s, 1H, Ar); ^{13}C-NMR (100 MHz, CDCl$_3$) δ: 20.3, 42.3 (2C), 118.8, 124.5, 145.9, 148.5, 152.4, 153.0; *Anal.* calcd for C$_9$H$_{10}$BrN$_3$: C, 45.02; H, 4.20; N, 17.50. Found: C, 44.74; H, 4.13; N, 17.43.

2.2.1.11 2-(2-Bromopyridin-3-yl)-1,4,5,6-tetrahydropyrimidine (14)

2-Bromonicotinaldehyde (0.93 g, 5.0 mmol) was subjected to the general procedure. Yellow solid (1.14 g, 95 %): mp 106–108 °C (from CHCl$_3$–n-hexane); IR
(neat) cm^{-1}: 1626 (C=N); ^{1}H-NMR (400 MHz, CDCl$_3$) δ: 1.84–1.89 (m, 2H,
CH$_2$), 3.44 (t, J = 5.9 Hz, 4H, 2 × CH$_2$), 4.89 (br s, 1H, NH), 7.27–7.30 (m, 1H,
Ar), 7.72 (dd, J = 7.6, 2.0 Hz, 1H, Ar), 8.35 (d, J = 4.8, 2.0 Hz, 1H, Ar); ^{13}C-
NMR (100 MHz, CDCl$_3$) δ: 20.3, 42.3 (2C), 122.7, 136.1, 138.7, 140.1, 150.1,
153.8; HRMS (FAB): m/z calcd for C$_9$H$_{11}$BrN$_3$ [M + H]$^+$ 240.0136; found:
240.0139.

2.2.1.12 2-(1-Bromonaphthalen-2-yl)-1,4,5,6-tetrahydropyrimidine (16)

1-Bromo-2-naphthaldehyde (0.94 g, 4.0 mmol) was subjected to the general procedure. Colorless crystals (1.04 g, 90 %): mp 151–153 °C (from CHCl$_3$–n-hex-
ane); IR (neat) cm^{-1}: 1625 (C=N); ^{1}H-NMR (400 MHz, CDCl$_3$) δ: 1.89–1.95 (m,
2H, CH$_2$), 3.52 (t, J = 5.7 Hz, 4H, 2 × CH$_2$), 4.72 (br s, 1H, NH), 7.47-7.62 (m,
3H, Ar), 7.77–7.82 (m, 2H, Ar), 8.33 (d, J = 8.5 Hz, 1H, Ar); ^{13}C-NMR
(100 MHz, CDCl$_3$) δ: 20.7, 42.5 (2C), 121.2, 126.8, 126.9, 127.6, 127.6, 127.9,
128.1, 132.1, 134.2, 137.5, 156.1; $Anal.$ calcd for C$_{14}$H$_{13}$BrN$_2$: C, 58.15; H, 4.53;
N, 9.69. Found: C, 58.02; H, 4.47; N, 9.71.

2.2.1.13 2-(2-Bromophenyl)-4,5-dihydro-1H-imidazole (24)

2-Bromobenzaldehyde (0.93 g, 5.0 mmol) was subjected to the general procedure
using ethylenediamine (0.33 g, 5.5 mmol) instead of propylenediamine. Colorless
crystals (0.77 g, 68 %): mp 98–99 °C (from CHCl$_3$–n-hexane); IR (neat) cm^{-1}:
1619 (C=N); ^{1}H-NMR (500 MHz, CDCl$_3$) δ: 3.79 (s, 4H, 2 × CH$_2$), 4.99 (br s,
1H, NH), 7.25 (ddd, J = 8.0, 7.5, 1.7 Hz, 1H, Ar), 7.33 (ddd, J = 8.0, 7.5, 1.1 Hz,
1H, Ar), 7.58 (dd, J = 8.0, 1.1 Hz, 1H, Ar), 7.64 (d, J = 8.0, 1.7 Hz, 1H, Ar);
^{13}C-NMR (125 MHz, CDCl$_3$) δ: 50.6 (2C), 120.8, 127.4, 131.0, 131.2, 133.0,
133.2, 164.4; HRMS (FAB): m/z calcd for C$_9$H$_{10}$BrN$_2$ [M + H]$^+$ 225.0027; found:
225.0030.

2.2.1.14 General Procedure for Cyclization Using CS$_2$. 3,4-Dihydro-2H,6H-pyrimido[1,2-c][1,3]benzo- thiazine-6-thione (7a)

To a mixture of **6aa** (59.8 mg, 0.25 mmol) and NaH (20.0 mg, 0.50 mmol; 60 %
oil suspension) in DMF (0.83 mL) was added CS$_2$ (30.5 μL, 0.50 mmol) under an
Ar atmosphere. After being stirred at 80 °C for 12 h, the mixture was concentrated
in $vacuo$. The residue was purified by flash chromatography over silica gel with n-

hexane–EtOAc (9:1) to give the compound **7a** as a pale-yellow solid (51.4 mg, 88 %). Spectral data were in good agreement with compound **24** in Chap. 2.

2.2.1.15 3,4-Dihydro-9-methoxy-2*H*,6*H*-pyrimido[1,2-*c*][1,3]benzothiazine-6-thione (7b)

The fluoride **6b** (52.1 mg, 0.25 mmol) was subjected to the general procedure. Pale yellow solid (62.6 mg, 95 %): mp 120–122 °C (from CHCl$_3$–*n*-hexane); IR (neat) cm^{-1}: 1624 (C=N); ^{1}H-NMR (500 MHz, CDCl$_3$) δ: 2.00–2.05 (m, 2H, CH$_2$), 3.71 (t, J = 5.4 Hz, 2H, CH$_2$), 3.83 (m, 3H, OCH$_3$), 4.42 (t, J = 6.3 Hz, 2H, CH$_2$), 6.46 (d, J = 2.3 Hz, 1H, Ar), 6.83 (dd, J = 9.0, 2.3 Hz, 1H, Ar), 8.12 (d, J = 9.0 Hz, 1H, Ar); ^{13}C-NMR (125 MHz, CDCl$_3$) δ: 21.6, 45.3, 48.7, 55.6, 104.9, 114.9, 119.2, 130.7, 133.3, 143.9, 161.6, 189.7; *Anal.* calcd for C$_{12}$H$_{12}$N$_2$OS$_2$: C, 54.52; H, 4.58; N, 10.60. Found: C, 54.22; H, 4.62; N, 10.47.

2.2.1.16 3,4-Dihydro-9-methyl-2*H*,6*H*-pyrimido[1,2-*c*][1,3]benzothiazine-6-thione (7c)

The bromide **6c** (63.3 mg, 0.25 mmol) was subjected to the general procedure. Colorless solid (54.6 mg, 88 %): mp 146–147 °C (from CHCl$_3$–*n*-hexane); IR (neat) cm^{-1}: 1620 (C=N); ^{1}H-NMR (400 MHz, CDCl$_3$) δ: 1.99–2.06 (m, 2H, CH$_2$), 2.35 (m, 3H, CH$_3$), 3.73 (t, J = 5.6 Hz, 2H, CH$_2$), 4.43 (t, J = 6.2 Hz, 2H, CH$_2$), 6.82 (d, J = 1.0 Hz, 1H, Ar), 7.10 (dd, J = 8.3, 1.0 Hz, 1H, Ar), 8.07 (d, J = 8.3 Hz, 1H, Ar); ^{13}C-NMR (100 MHz, CDCl$_3$) δ: 21.2, 21.6, 45.4, 48.7, 121.6, 123.8, 128.7, 128.7, 131.6, 141.8, 144.2, 190.0; *Anal.* calcd for C$_{12}$H$_{12}$N$_2$S$_2$: C, 58.03; H, 4.87; N, 11.28. Found: C, 57.84; H, 4.85; N, 11.19.

2.2.1.17 9-Fluoro-3,4-dihydro-2*H*,6*H*-pyrimido[1,2-*c*][1,3]benzothiazine-6-thione (7d)

The bromide **6d** (64.3 mg, 0.25 mmol) was subjected to the general procedure. Colorless solid (47.9 mg, 76 %): mp 185 °C (from CHCl$_3$–*n*-hexane); IR (neat) cm^{-1}: 1630 (C=N); ^{1}H-NMR (500 MHz, CDCl$_3$) δ: 2.01–2.06 (m, 2H, CH$_2$), 3.73 (t, J = 5.7 Hz, 2H, CH$_2$), 4.42 (t, J = 6.0 Hz, 2H, CH$_2$), 6.73 (dd, J = 8.0, 2.9 Hz, 1H, Ar), 6.98 (ddd, J = 8.9, 8.9, 2.9 Hz, 1H, Ar), 8.22 (dd, J = 8.9, 5.4 Hz, 1H, Ar); ^{13}C-NMR (125 MHz, CDCl$_3$) δ: 21.5, 45.4, 48.7, 108.1 (d, J = 24.0 Hz), 115.1 (d, J = 22.8 Hz), 122.7 (d, J = 3.6 Hz), 131.7 (d, J = 9.6 Hz), 134.0 (d, J = 8.4 Hz), 143.4, 163.9 (d, J = 255.5 Hz), 188.9; ^{19}F-NMR (500 MHz, CDCl$_3$) δ: –106.9; *Anal.* calcd for C$_{11}$H$_9$FN$_2$S$_2$: C, 52.36; H, 3.60; N, 11.10. Found: C, 52.10; H, 3.48; N, 11.15.

2.2.1.18 3,4-Dihydro-9-nitro-2*H*,6*H*-pyrimido[1,2-c][1,3]benzothiazine-6-thione (7e)

Using the general procedure, the fluoride **6e** (55.8 mg, 0.25 mmol) was allowed to react with CS$_2$ at rt for 12 h. Pale yellow solid (50.7 mg, 73 %): mp 192–193 °C (from CHCl$_3$–*n*-hexane); IR (neat) cm^{-1}: 1620 (C=N), 1598 (NO$_2$), 1520 (NO$_2$); ^{1}H-NMR (500 MHz, CDCl$_3$) δ: 2.05-2.09 (m, 2H, CH$_2$), 3.81 (t, *J* = 5.7 Hz, 2H, CH$_2$), 4.44 (t, *J* = 6.0 Hz, 2H, CH$_2$), 7.90 (d, *J* = 1.7 Hz, 1H, Ar), 8.06 (dd, *J* = 9.0, 1.7 Hz, 1H, Ar), 8.40 (d, *J* = 9.0 Hz, 1H, Ar); ^{13}C-NMR (125 MHz, CDCl$_3$) δ: 21.4, 45.8, 48.5, 117.0, 121.4, 130.7, 131.2, 133.8, 142.9, 149.0, 187.9; *Anal.* calcd for C$_{11}$H$_9$N$_3$O$_2$S$_2$: C, 47.30; H, 3.25; N, 15.04. Found: C, 47.07; H, 3.19; N, 14.99.

2.2.1.19 3,4-Dihydro-10-methoxy-2*H*,6*H*-pyrimido[1,2-c][1,3]benzothiazine-6-thione (7f)

The bromide **6f** (67.3 mg, 0.25 mmol) was subjected to the general procedure. Pale yellow solid (11.3 mg, 17 %): mp 136 °C (from CHCl$_3$–*n*-hexane); IR (neat) cm^{-1}: 1625 (C=N); ^{1}H-NMR (400 MHz, CDCl$_3$) δ: 1.99–2.07 (m, 2H, CH$_2$), 3.76 (t, *J* = 5.7 Hz, 2H, CH$_2$), 3.86 (s, 3H, OCH$_3$), 4.46 (t, *J* = 6.2 Hz, 2H, CH$_2$), 6.94 (d, *J* = 8.8 Hz, 1H, Ar), 7.02 (dd, *J* = 8.8, 2.7 Hz, 1H, Ar), 7.75 (d, *J* = 2.7 Hz, 1H, Ar); ^{13}C-NMR (100 MHz, CDCl$_3$) δ: 21.6, 45.5, 48.6, 55.6, 111.6, 119.8, 123.0, 123.4, 127.6, 144.2, 159.2, 189.9; HRMS (FAB): *m/z* calcd for C$_{12}$H$_{13}$N$_2$OS$_2$ [M + H]$^+$ 265.0469; found: 265.0461.

2.2.1.20 3,4-Dihydro-10-nitro-2*H*,6*H*-pyrimido[1,2-c][1,3]benzothiazine-6-thione (7g)

Using the general procedure, the bromide **6g** (71.0 mg, 0.25 mmol) was allowed to react with CS$_2$ at rt for 12 h. Yellow solid (39.6 mg, 57 %): mp 176–177 °C (from CHCl$_3$–*n*-hexane); IR (neat) cm^{-1}: 1627 (C=N), 1605 (NO$_2$), 1523 (NO$_2$); ^{1}H-NMR (500 MHz, CDCl$_3$) δ: 2.05-2.10 (m, 2H, CH$_2$), 3.81 (t, *J* = 5.4 Hz, 2H, CH$_2$), 4.44 (t, *J* = 6.0 Hz, 2H, CH$_2$), 7.18 (d, *J* = 8.9 Hz, 1H, Ar), 8.23 (dd, *J* = 8.9, 2.9 Hz, 1H, Ar), 9.09 (d, *J* = 2.9 Hz, 1H, Ar); ^{13}C-NMR (125 MHz, CDCl$_3$) δ: 21.3, 45.6, 48.6, 122.7, 124.6, 125.4, 127.4, 139.2, 142.4, 146.9, 187.3; HRMS (FAB): *m/z* calcd for C$_{11}$H$_{10}$N$_3$O$_2$S$_2$ [M + H]$^+$ 280.0214; found: 280.0211.

2.2.1.21 3,4-Dihydro-2*H*,6*H*-pyrimido[1,2-*c*]pyrido[4,3-*e*][1,3]thiazine-6-thione (13)

The bromide **12** (60.0 mg, 0.25 mmol) was subjected to the general procedure. Orange solid (10.8 mg, 18 %): mp 205–207 °C (from CHCl$_3$–*n*-hexane); IR (neat) cm^{-1}: 1619 (C=N); ^{1}H-NMR (400 MHz, CDCl$_3$) δ: 2.03–2.09 (m, 2H, CH$_2$), 3.79 (t, J = 5.6 Hz, 2H, CH$_2$), 4.44 (t, J = 6.1 Hz, 2H, CH$_2$), 8.01 (d, J = 5.4 Hz, 1H, Ar), 8.36 (s, 1H, Ar), 8.51 (d, J = 5.4 Hz, 1H, Ar); ^{13}C-NMR (100 MHz, CDCl$_3$) δ: 21.4, 45.7, 48.4, 121.3, 128.2, 132.9, 142.6, 143.2, 148.3, 188.5; HRMS (FAB): *m/z* calcd for C$_{10}$H$_{10}$N$_3$S$_2$ [M + H]$^+$ 236.0316; found: 236.0311.

2.2.1.22 3,4-Dihydro-2*H*,6*H*-pyrimido[1,2-*c*]pyrido[3,2-*e*][1,3]thiazine-6-thione (15)

The bromide **14** (60.0 mg, 0.25 mmol) was subjected to the general procedure. Colorless solid (41.9 mg, 71 %): mp 141–142 °C (from CHCl$_3$–*n*-hexane); IR (neat) cm-1: 1621 (C=N); ^{1}H-NMR (400 MHz, CDCl$_3$) δ: 2.02–2.08 (m, 2H, CH$_2$), 3.76 (t, J = 5.6 Hz, 2H, CH$_2$), 4.45 (t, J = 6.2 Hz, 2H, CH$_2$), 7.22 (dd, J = 8.0, 4.5 Hz, 1H, Ar), 8.46 (dd, J = 8.0, 1.6 Hz, 1H, Ar), 8.54 (dd, J = 4.5, 1.6 Hz, 1H, Ar); ^{13}C-NMR (100 MHz, CDCl$_3$) δ: 21.3, 45.7, 48.6, 122.2, 124.0, 136.5, 143.8, 151.9, 153.3, 190.8; *Anal.* calcd for C$_{10}$H$_9$N$_3$S$_2$: C, 51.04; H, 3.85; N, 17.86. Found: C, 50.88; H, 3.95; N, 17.82.

2.2.1.23 2,3-Dihydronaphtho[2,1-*e*]pyrimido[1,2-*c*][1,3]thiazine-12(1*H*)-thione (17)

The bromide **16** (72.3 mg, 0.25 mmol) was subjected to the general procedure. Pale yellow solid (73.4 mg, >99 %): mp 230–231 °C (from CHCl$_3$–*n*-hexane); IR (neat) cm^{-1}: 1620 (C=N); ^{1}H-NMR (400 MHz, CDCl$_3$) δ: 2.06-2.12 (m, 2H, CH$_2$), 3.82 (t, J = 5.5 Hz, 2H, CH$_2$), 4.50 (t, J = 6.2 Hz, 2H, CH$_2$), 7.58-7.63 (m, 2H, Ar), 7.75 (d, J = 9.0 Hz, 1H, Ar), 7.83-7.86 (m, 1H, Ar), 7.96-8.00 (m, 1H, Ar), 8.26 (d, J = 8.8 Hz, 1H, Ar); ^{13}C-NMR (100 MHz, CDCl$_3$) δ: 21.5, 45.7, 48.7, 123.0, 124.0, 124.7, 126.0, 127.1, 127.3, 128.3, 128.4, 129.7, 133.9, 144.8, 188.4; *Anal.* calcd for C$_{15}$H$_{12}$N$_2$S$_2$: C, 63.35; H, 4.25; N, 9.85. Found: C, 63.36; H, 4.03; N, 9.70.

2.2.1.24 General Procedure for Cyclization Using Isothiocyanates or Isocyanates. *N*-Benzyl-3,4-dihydro-2*H*-pyrimido[1,2-*c*]quinazolin-6(7*H*)-thione (18)

To a mixture of the fluoride **6ab** (44.6 mg, 0.25 mmol) and NaH (20.0 mg, 0.50 mmol; 60 % oil suspension) in DMF (0.83 mL) was added benzylisothiocyanate (66.0 μL, 0.50 mmol) under an Ar atmosphere. After being stirred at rt for 2 h, EtOAc was added. The resulting solution was washed with sat. NaHCO$_3$, brine, and dried over Na$_2$SO$_4$. After concentration, the residue was purified by flash chromatography over aluminum oxide with *n*-hexane–EtOAc (29:1) to give the title compound **18** as a colorless solid (74.7 mg, 97 %): mp 137 °C (from CHCl$_3$–*n*-hexane); IR (neat) cm^{-1}: 1635 (C=N); ^{1}H-NMR (500 MHz, CDCl$_3$) δ: 2.02-2.07 (m, 2H, CH$_2$), 3.67 (t, *J* = 5.2 Hz, 2H, CH$_2$), 4.43 (t, *J* = 6.0 Hz, 2H, CH$_2$), 6.01 (br s, 2H, CH$_2$), 6.99 (d, *J* = 8.6 Hz, 1H, Ar), 7.15–7.36 (m, 7H, Ar), 8.18 (d, *J* = 7.4 Hz, 1H, Ar); ^{13}C-NMR (125 MHz, CDCl$_3$) δ: 21.7, 44.8, 50.3, 54.5, 115.6, 119.6, 124.4, 126.0, 126.2 (2C), 127.2, 128.8 (2C), 132.2, 135.6, 137.5, 143.2, 177.6; *Anal.* calcd for C$_{18}$H$_{17}$N$_3$S: C, 70.33; H, 5.57; N, 13.67. Found: C, 70.31; H, 5.66; N, 13.69.

2.2.1.25 *N*-(*tert*-Butyl)-3,4-dihydro-2*H*,6*H*-pyrimido[1,2-c][1,3]benzothiazin-6-imine (19)

Using the general procedure, the fluoride **6ab** (44.6 mg, 0.25 mmol) was allowed to react with *tert*-butylisothiocyanate (63.4 μL, 0.50 mmol) at 80 °C for 2 h and purified by flash chromatography over aluminum oxide with *n*-hexane–EtOAc (1:0 to 9:1). Pale yellow solid (42.7 mg, 62 %): mp 62 °C (from *n*-hexane): IR (neat) cm^{-1}: 1622 (C=N), 1598 (C=N); ^{1}H-NMR (400 MHz, CDCl$_3$) δ: 1.39 (s, 9H, 3 × CH$_3$), 1.88–1.94 (m, 2H, CH$_2$), 3.62 (t, *J* = 5.6 Hz, 2H, CH$_2$), 3.87 (t, *J* = 6.2 Hz, 2H, CH$_2$), 7.11 (dd, *J* = 8.0, 1.2 Hz, 1H, Ar), 7.20 (ddd, *J* = 8.0, 7.3, 1.2 Hz, 1H, Ar), 7.31 (ddd, *J* = 8.0, 7.3, 1.4 Hz, 1H, Ar), 8.18 (dd, *J* = 8.0, 1.4 Hz, 1H, Ar); ^{13}C-NMR (125 MHz, CDCl$_3$) δ: 21.9, 30.0 (3C), 45.1, 45.4, 54.1, 124.5, 126.0, 127.8, 128.4, 129.0, 130.1, 138.3, 148.0; HRMS (FAB): *m/z* calcd for C$_{15}$H$_{20}$N$_3$S [M + H]$^+$ 274.1378; found: 274.1375.

2.2.1.26 *N*-Benzyl-3,4-dihydro-2*H*-pyrimido[1,2-c]quinazolin-6-one (20)

Using the general procedure, the fluoride **6ab** (44.6 mg, 0.25 mmol) was allowed to react with benzylisocyanate (61.6 μL, 0.50 mmol) at rt for 2 h and purified by flash chromatography over aluminum oxide with *n*-hexane–EtOAc (1:0 to 9:1). White solid (74.8 mg, >99 %): mp 105–107 °C (from CHCl$_3$–*n*-hexane); IR (neat) cm^{-1}: 1672 (C=O), 1625 (C=N); ^{1}H-NMR (400 MHz, CDCl$_3$) δ: 1.97–2.03 (m, 2H, CH$_2$), 3.67 (t, *J* = 5.5 Hz, 2H, CH$_2$), 3.99 (t, *J* = 6.0 Hz, 2H, CH$_2$), 5.29 (s, 2H, CH$_2$), 6.93 (d, *J* = 8.3 Hz, 1H, Ar), 7.09 (dd, *J* = 8.0, 7.3 Hz, 1H, Ar), 7.23–7.35 (m, 6H, Ar), 8.18 (d, *J* = 8.0 Hz, 1H, Ar); ^{13}C-NMR (100 MHz, CDCl$_3$) δ: 20.6, 41.8, 44.5, 47.0, 114.1, 117.8, 122.7, 126.1, 126.4, 127.3, 127.3, 128.6, 128.8, 132.0, 136.4, 138.0, 145.3, 150.8; *Anal.* calcd for C$_{18}$H$_{17}$N$_3$O: C, 74.20; H, 5.88; N, 14.42. Found: C, 73.90; H, 6.04; N, 14.12.

2.2.1.27 N-(tert-Butyl)-3,4-dihydro-2H-pyrimido[1,2-c]quinazolin-6-one (21) and N-(tert-Butyl)-3,4-dihydro-2H- pyrimido [1,2-c][1,3]benzoxazin-6-imine (22)

Using the general procedure, the fluoride **6ab** (44.6 mg, 0.25 mmol) was allowed to react with *tert*-butylisocyanate (57.1 µL, 0.50 mmol) at 80 °C for 2 h and purified by flash chromatography over aluminum oxide with *n*-hexane–EtOAc (1:0 to 9:1).

Compound **21**: pale yellow oil (34.9 mg, 54 %): IR (neat) cm^{-1}: 1679 (C=O), 1631 (C=N); ^{1}H-NMR (400 MHz, CDCl$_3$) δ: 1.68 (s, 9H, 3 × CH$_3$), 1.86–1.92 (m, 2H, CH$_2$), 3.61 (t, J = 5.5 Hz, 2H, CH$_2$), 3.77 (t, J = 6.2 Hz, 2H, CH$_2$), 7.07-7.11 (m, 1H, Ar), 7.25–7.27 (m, 1H, Ar), 7.33-7.37 (m, 1H, Ar), 7.95 (dd, J = 7.8, 1.2 Hz, 1H, Ar); ^{13}C-NMR (100 MHz, CDCl$_3$) δ: 20.9, 30.4 (3C), 41.3, 44.6, 59.6, 119.5, 122.5, 122.6, 125.9, 129.3, 138.8, 147.3, 151.7; HRMS (FAB): *m/z* calcd for C$_{15}$H$_{20}$N$_3$O [M + H]$^+$ 258.1606; found: 258.1604.

Compound **22**: colorless crystals (11.4 mg, 18 %): mp 53–55 °C (from CHCl$_3$–*n*-hexane); IR (neat) cm^{-1}: 1637 (C=N), 1613 (C=N); ^{1}H-NMR (500 MHz, CDCl$_3$) δ: 1.35 (s, 9H, 3 × CH$_3$), 1.91–1.96 (m, 2H, CH$_2$), 3.59 (t, J = 5.7 Hz, 2H, CH$_2$), 3.79 (t, J = 6.0 Hz, 2H, CH$_2$), 7.01 (d, J = 8.0 Hz, 1H, Ar), 7.12 (dd, J = 7.7, 7.4 Hz, 1H, Ar), 7.40 (ddd, J = 8.0, 7.4, 1.4 Hz, 1H, Ar), 8.00 (dd, J = 7.7, 1.4 Hz, 1H, Ar); ^{13}C-NMR (125 MHz, CDCl$_3$) δ: 21.0, 30.8 (3C), 43.4, 44.3, 52.5, 115.1, 116.6, 123.5, 125.5, 132.0, 139.1, 143.8, 150.6; HRMS (FAB): *m/z* calcd for C$_{15}$H$_{20}$N$_3$O [M + H]$^+$ 258.1606; found: 258.1602.

2.2.1.28 N-Phenyl-3,4-dihydro-2H-pyrimido[1,2-c]quinazolin-6-one (23)

Using the general procedure, the fluoride **6ab** (44.6 mg, 0.25 mmol) was allowed to react with phenylisocyanate (54.5 µL, 0.50 mmol) at rt for 2 h and purified by flash chromatography over aluminum oxide with *n*-hexane–EtOAc (1:0 to 9:1). White solid (69.6 mg, >99 %): mp 225–226 °C (from CHCl$_3$–*n*-hexane); IR (neat) cm^{-1}: 1684 (C=O), 1629 (C=N); ^{1}H-NMR (400 MHz, CDCl$_3$) δ: 1.97-2.03 (m, 2H, CH$_2$), 3.69 (t, J = 5.6 Hz, 2H, CH$_2$), 3.95 (t, J = 6.0 Hz, 2H, CH$_2$), 6.37 (d, J = 8.3 Hz, 1H, Ar), 7.08–7.12 (m, 1H, Ar), 7.22–7.34 (m, 3H, Ar), 7.46-7.61 (m, 3H, Ar), 8.19 (dd, J = 7.9, 1.6 Hz, 1H, Ar); ^{13}C-NMR (100 MHz, CDCl$_3$) δ: 20.5, 41.5, 44.6, 115.1, 117.2, 122.8, 125.9, 128.8, 129.3 (2C), 130.1 (2C), 131.6, 137.2, 139.6, 145.4, 150.2; *Anal.* calcd for C$_{17}$H$_{15}$N$_3$O: C, 73.63; H, 5.45; N, 15.15. Found: C, 73.41; H, 5.27; N, 15.11.

2.2.1.29 N-(tert-Butyl)-2,3-dihydroimidazo[1,2-c][1,3]benzothiazin-5-imine (25)

Using the general procedure, the bromide **24** (112.5 mg, 0.50 mmol) was allowed to react with *tert*-butylisothiocyanate (126.8 μL, 1.00 mmol) at rt for 3 h and purified by flash chromatography over aluminum oxide with *n*-hexane–EtOAc (1:0 to 9:1). White solid (63.0 mg, 49 %): mp 140–142 °C (from CHCl$_3$–*n*-hexane); IR (neat) cm^{-1}: 1625 (C=N), 1604 (C=N); ^{1}H-NMR (500 MHz, CDCl$_3$) δ: 1.37 (s, 9H, 3 × CH$_3$), 3.93–4.03 (m, 4H, 2 × CH$_2$), 7.17 (d, J = 8.0 Hz, 1H, Ar), 7.22 (dd, J = 8.0, 7.7 Hz, 1H, Ar), 7.37 (dd, J = 7.7, 7.4 Hz, 1H, Ar), 8.19 (d, J = 7.4 Hz, 1H, Ar); ^{13}C-NMR (125 MHz, CDCl$_3$) δ: 30.1 (3C), 49.2, 52.2, 53.9, 121.8, 124.5, 126.0, 128.5, 131.5, 132.9, 134.6, 154.5; HRMS (FAB): *m/z* calcd for C$_{14}$H$_{18}$N$_3$S [M + H]$^+$ 260.1221; found: 260.1219.

2.2.1.30 3,4-Dihydro-2H,6H-pyrimido[1,2-c][1,3]benzothiazin-6-imine (26)

Synthesis from **7a**: 0.1 M solution of NaOH in a mixed solvent of MeOH/H$_2$O (9:1; 5 mL) was added to a flask containing **7a** (58.6 mg, 0.25 mmol). After being stirred under reflux for 12 h, the mixture was concentrated *in vacuo* [azeotroped with MeOH (×2) and CHCl$_3$ (×2)]. The residue was suspended with anhydrous EtOH (1 mL), and BrCN (53.0 mg, 0.50 mmol) was added under an Ar atmosphere. After being stirred under reflux for 2 h, 2 N NaOH was added to the mixture. The whole was extracted with CHCl$_3$, and dried over Na$_2$SO$_4$. After concentration, the residue was purified by flash chromatography over aluminum oxide with *n*-hexane–EtOAc (9:1) to give the title compound **26** as white solid (33.2 mg, 61 % in 2 steps).

Synthesis from **19**: TFA (2 mL) was added to a mixture of **19** (54.7 mg, 0.20 mmol) in small amount of CHCl$_3$ (1 or 2 drops) and MS4Å (300 mg, powder, activated by heating with Bunsen burner). After being stirred under reflux for 1 h, the mixture was concentrated *in vacuo*. To a stirring mixture of this residue in CHCl$_3$ was added dropwise Et$_3$N at 0 °C to adjust pH to 8–9. The whole was extracted with EtOAc. The extract was washed with sat. NaHCO$_3$ (×2), brine, and dried over Na$_2$SO$_4$. After concentration, the residue was purified by flash chromatography over aluminum oxide with *n*-hexane–EtOAc (9:1) to give the title compound **26** as white solid (36.9 mg, 85 %).

Compound **26**: mp 105 °C (from CHCl$_3$–*n*-hexane); IR (neat) cm^{-1}: 1621 (C=N), 1578 (C=N); ^{1}H-NMR (500 MHz, CDCl$_3$) δ: 1.95–2.00 (m, 2H, CH$_2$), 3.69 (t, J = 5.5 Hz, 2H, CH$_2$), 4.02 (t, J = 6.1 Hz, 2H, CH$_2$), 7.04 (dd, J = 7.5, 1.1 Hz, 1H, Ar), 7.17 (br s, 1H, NH), 7.21–7.24 (m, 1H, Ar), 7.34 (ddd, J = 7.5, 7.5, 1.4 Hz, 1H, Ar), 8.22 (dd, J = 7.5, 1.4 Hz, 1H, Ar); ^{13}C-NMR (100 MHz, CDCl$_3$) δ: 21.1, 43.8, 44.9, 123.5, 126.2, 126.8, 128.8, 128.9, 130.5, 146.6, 153.4; *Anal.* calcd for C$_{11}$H$_{11}$N$_3$S: C, 60.80; H, 5.10; N, 19.34.

References

1. Helmut, H., Juergen, M., Hans, Z.: Eur. Pat. Appl. (1982) EP 43936
2. Brown, J.P.: J. Chem. Soc., Perkin Trans. 1 869–870 (**1974**)
3. Peter, S.; Gerhard, S. Ger. Offen. (1979) DE 2811131
4. Schreiber, S.L.: Science **287**, 1964–1969 (2000)
5. Burke, M.D., Schreiber, S.L.: Angew. Chem. Int. Ed. **43**, 46–58 (2004)
6. Ishihara, M., Togo, H.: Tetrahedron **63**, 1474–1480 (2007)
7. Dyker, G.: Angew. Chem. Int. Ed. **38**, 1698–1712 (1999)
8. Ritleng, V., Sirlin, C., Pfeffer, M.: Chem. Rev. **102**, 1731–1770 (2002)
9. Alberico, D., Scott, M.E., Lautens, M.: Chem. Rev. **107**, 174–238 (2007)
10. Chen, H., Schlecht, S., Semple, T.C., Hartwig, J.F.: Science **287**, 1995–1997 (2000)
11. Dick, A.R., Hull, K.L., Sanford, M.S.J.: Am. Chem. Soc. **126**, 2300–2301 (2004)
12. Yu, J.-Q., Giri, R., Chen, X.: Org. Biomol. Chem. **4**, 4047 (2006)
13. Desai, L.V., Stowers, K.J., Sanford, M.S.J.: Am. Chem. Soc. **130**, 13285–13293 (2008)
14. Chatani, N., Asaumi, T., Yorimitsu, S., Ikeda, T., Kakiuchi, F., Murai, S.J.: Am. Chem. Soc. **123**, 10935–10941 (2001)
15. Zaitsev, V.G., Shabashov, D., Daugulis, O.J.: Am. Chem. Soc. **127**, 13154–13155 (2005)
16. Berman, A.M., Lewis, J.C., Bergman, R.G., Ellman, J.A.J.: Am. Chem. Soc. **130**, 14926–14927 (2008)
17. Williams, N.A., Uchimura, Y., Tanaka, M.: J. Chem. Soc., Chem. Commun. 1129–1130 (1995)
18. Kakiuchi, F., Yamauchi, M., Chatani, N., Murai, S.: Chem. Lett. 111–112 (1996)
19. Pastine, S.J., Gribkov, D.V., Sames, D.J.: Am. Chem. Soc. **128**, 14220–14221 (2006)
20. Kakiuchi, F., Sato, T., Yamauchi, M., Chatani, N., Murai, S.: Chem. Lett. 19–20 (1999)
21. Chen, X., Li, J–.J., Hao, X.-S., Goodhue, C.E., Yu, J.-Q.J.: Am. Chem. Soc. **128**, 78–79 (2006)
22. Oi, S., Aizawa, E., Ogino, Y., Inoue, Y.J.: Org. Chem. **70**, 3113–3119 (2005)
23. For a review on Cu-mediated C–H functionalization, see: Wendlandt, A.E., Suess, A.M., Stahl, S.S.: Angew. Chem., Int. Ed. **50**, 11062–11087 (2011)
24. Reinaud, O., Capdevielle, P., Maumy, M.: Synthesis 612–614 (1990)
25. Chen, X., Hao, X.-S., Goodhue, C.E., Yu, J.-Q.J.: Am. Chem. Soc. **128**, 6790–6791 (2006)
26. Uemura, T., Imoto, S., Chatani, N.: Chem. Lett. **35**, 842–843 (2006)
27. Brasche, G., Buchwald, S.L.: Angew. Chem. Int. Ed. **47**, 1932–1934 (2008)
28. Kochi, J.K., Tang, R.T., Bernath, T.J.: Am. Chem. Soc. **95**, 7114–7123 (1973)
29. Ramsden, C.A., Rose, H.L.: J. Chem. Soc., Perkin Trans. 1 615–617 (1995)
30. Ramsden, C.A., Rose, H.L.: J. Chem. Soc., Perkin Trans. 1 2319–2327 (1997)
31. Hickman, A.J., Sanford, M.S.: Nature **484**, 177–185 (2012)
32. Ribas, X., Jackson, D.A., Donnadieu, B., Mahía, J., Parella, T., Xifra, R., Hedman, B., Hodgson, K.O., Llobet, A., Stack, T.D.P.: Angew. Chem. Int. Ed. **41**, 2991–2994 (2002)
33. Xifra, R., Ribas, X., Llobet, A., Poater, A., Duran, M., Solà, M., Stack, T.D.P., Benet-Buchholz, J., Donnadieu, B., Mahía, J., Parella, T.: Chem. Eur. J. **11**, 5146–5156 (2005)
34. Huffman, L.M., Stahl, S.S.J.: Am. Chem. Soc. **130**, 9196–9197 (2008)
35. King, A.E., Brunold, T.C., Stahl, S.S.J.: Am. Chem. Soc. **131**, 5044–5045 (2009)
36. King, A.E., Huffman, L.M., Casitas, A., Costas, M., Ribas, X., Stahl, S.S.: J. Am. Chem. Soc. **132**, 12068–12073 (2010)
37. John, A., Nicholas, M.J.: Org. Chem. **76**, 4158–4162 (2011)
38. Hartwig, J.F.: Synlett 329–340 (1997)
39. Hartwig, J.F.: Angew. Chem. Int. Ed. **37**, 2046–2067 (1998)
40. Wolfe, J.P., Wagaw, S., Marcoux, J.F., Buchwald, S.L.: Acc. Chem. Res. **31**, 805–818 (1998)
41. Kunz, K., Scholz, U., Ganzer, D.: Synlett 2428–2439 (2003)
42. Ferraccioli, R., Carenzi, D.: Synthesis 1383–1386 (2003)
43. Orain, D., Blumstein, A.-C., Tasdelen, E., Haessig, S.: Synlett 2433–2436 (2008)

44. Shen, G., Lv, X., Bao, W.: Eur. J. Org. Chem. 5897–5901 (2009)
45. Qiu, J.-W., Zhang, X.-G., Tang, R.-Y., Zhong, P., Li, J.-H.: Adv. Synth. Catal. **351**, 2319–2323 (2009)
46. Murru, S., Ghosh, H., Sahoo, S.K., Patel, B.K.: Org. Lett. **11**, 4254–4257 (2009)
47. Murru, S., Mondal, P., Yella, R., Patel, B.K.: Eur. J. Org. Chem. 5406–5413 (2009)
48. Wang, F., Zhao, P., Xi, C.: Tetrahedron Lett. **52**, 231–235 (2011)
49. Kaname, M., Minoura, M., Sashida, H.: Tetrahedron Lett. **52**, 505–508 (2011)
50. Ma, D., Lu, X., Shi, L., Zhang, H., Jiang, Y., Liu, X.: Angew. Chem. Int. Ed. **50**, 1118–1121 (2011)
51. Yang, J., Li, P., Wang, L.: Tetrahedron **67**, 5543–5549 (2011)
52. Wang, F., Cai, S., Liao, Q., Xi, C.J.: Org. Chem. **76**, 3174–3180 (2011)
53. Lach, F.: Synlett 2639–2642 (2012)
54. D'Amico, J.J., Tung, C.C., Dahl, W.E., Dahm, D.J.J.: Org. Chem. **41**, 3564–3568 (1976)
55. Leymarie-Beljean, M., Pays, M., Richer, J.-C.J.: Heterocycl. Chem. **17**, 1175–1179 (1980)
56. Anderson-McKay, J.E., Liepa, A.J.: Aust. J. Chem. **40**, 1179–1190 (1987)
57. Kobayashi, K., Komatsu, T., Konishi, H.: Heterocycles **78**, 2559–2564 (2009)
58. Zhu, L., Zhang, M.J.: Org. Chem. **69**, 7371–7374 (2004)
59. Mizuhara, T., Oishi, S., Fujii, N., Ohno, H.J.: Org. Chem. **75**, 265–268 (2010)
60. Wang, F., Cai, F., Wang, Z., Xi, C.: Org. Lett. **13**, 3202–3205 (2011)
61. Zhao, P., Wang, F., Xi, C.: Synthesis 1477–1480 (2012)
62. Muthusamy, S., Paramasivam, R., Ramakrishnan, V.T.J.: Heterocycl. Chem. **28**, 759–763 (1991)
63. Zambounis, J.S., Christen, E., Pfeiffer, J., Ribs, G.J.: Am. Chem. Soc. **116**, 925–931 (1994)
64. Huang, S., Connolly, P.J.: Tetrahedron Lett. **45**, 9373–9375 (2004)
65. Bunnet, J.F., Zahler, R.E.: Chem. Rev. **49**, 273–412 (1951)
66. Buncel, E., Dust, J.M., Terrier, F.: Chem. Rev. **95**, 2261–2280 (1995)
67. Annulli, A., Mencarelli, P., Stegel, F.J.: Org. Chem. **49**, 4065–4067 (1984)
68. Gorvin, J. H.: J. Chem. Soc., Perkin Trans. 1 1331–1335 (1988)
69. Raeppel, S., Raeppel, F., Suffert, J.: Synlett 794–796 (1998)
70. Ratz, A.M., Weigel, L.O.: Tetrahedron Lett. **40**, 2239–2242 (1999)
71. Rogers, J.F., Green, D.M.: Tetrahedron Lett. **43**, 3585–3587 (2002)
72. Grecian, S.A., Hadida, S., Warren, S.D.: Tetrahedron Lett. **46**, 4683–4685 (2005)
73. Barbero, N., SanMartin, R., Domínguez, E.: Tetrahedron **65**, 5729–5732 (2009)

Chapter 3
Structure–Activity Relationship Study of PD 404182 Derivatives for the Highly Potent Anti-HIV Agents

3,4-Dihydro-2*H*,6*H*-pyrimido[1,2-*c*][1,3]benzothiazin-6-imine (PD 404182) (**1**, Fig. 3.1) is a promising antiviral agent because of its high therapeutic index (CC_{50}/ $EC_{50} > 200$) and broad spectrum antiviral activities including against hepatitis C virus (HCV), simian immunodeficiency virus (SIV), and vesicular stomatitis virus (VSV), as well as HIV [1, 2]. In this chapter, the author describes the structure–activity relationship (SAR) studies of PD 404182 for the development of highly potent anti-HIV agents using the novel synthetic methods.

PD 404182 consists of three components, namely a 1,3-thiazin-2-imine core, and left-fused benzene and cyclic amidine moieties (Fig. 3.2). In order to obtain detailed insights into the relationships between compound structure and anti-HIV activity, the author planned to investigate substituent effects on each component: (I) derivatives with various heteroatom (N, S, and O) arrangements on the 1,3-thiazin-2-imine core; (II) pyrimido[1,2-*c*][1,3]thiazin-6-imine derivatives fused with a substituted benzene ring or a five- or six-membered aromatic heterocycle; and (III) benzo[*e*][1,3]thiazin-2-imine derivatives fused with a cyclic amidine ring with or without accessory alkyl or aryl groups.

The investigation began with the synthesis of tricyclic heterocycles with different combinations of heteroatoms on the 1,3-thiazin-2-imine core. As described in Chap. 2, the author developed synthetic methods for pyrimido[1,2-*c*][1,3]benzoxazine, pyrimido[1,2-*c*] quinazoline, and pyrimido[1,2-*c*][1,3]benzothiazine derivatives using Cu(II)-mediated C–H functionalization. This facilitates the introduction of oxygen, nitrogen, and sulfur functional groups at the *ortho*-position of 2-phenyl-1,4,5,6-tetrahydropyrimidine (**2**).

One-pot reaction for Cu(OAc)$_2$-mediated C–H functionalization of **2** and subsequent treatment with triphosgene provided a 1,3-oxazin-2-one derivative **4** (Scheme 3.1). The same one-pot procedure using thiophosgene produced a trace amount of the desired thiocarbonyl derivative **5**. Treatment of the purified compound **3** with thiophosgene provided the desired 1,3-oxazin-2-thione **5** in high yield. 1,3-Oxazin-2-imine **6** was obtained by the reaction of **3** with BrCN.

The Cu-mediated C–N bond formation of compound **2** with *tert*-butylcarbamate followed by spontaneous intramolecular cyclization afforded a pyrimido[1,2-*c*]quinazolin-6-one scaffold **7** (Scheme 3.1). Subsequent treatment with

T. Mizuhara, *Development of Novel Anti-HIV Pyrimidobenzothiazine Derivatives*, Springer Theses, DOI: 10.1007/978-4-431-54445-6_3, © Springer Japan 2013

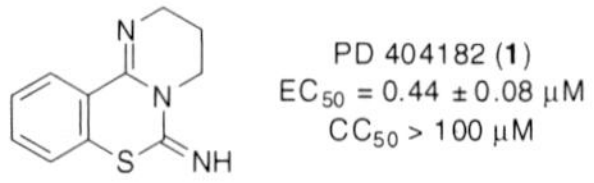

Fig. 3.1 Structure of PD 404182

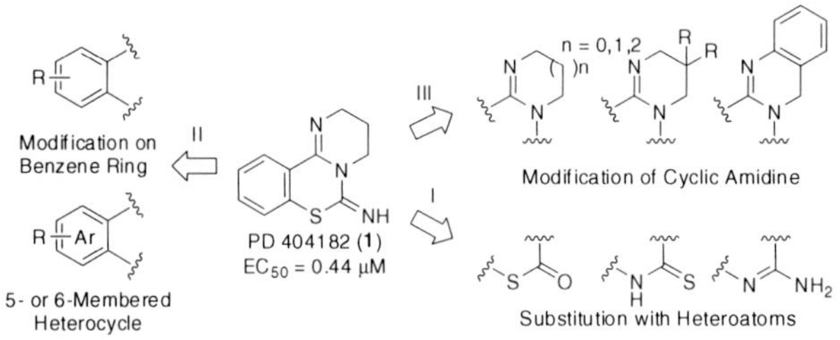

Fig. 3.2 Strategy for the SAR study of PD 404182

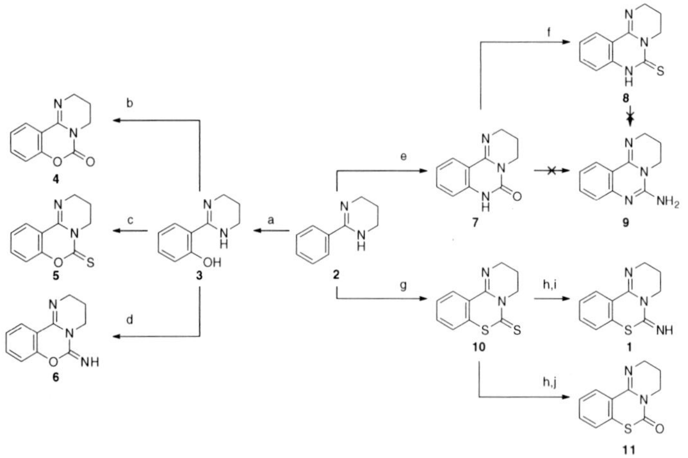

Scheme 3.1 Syntheses of various tricyclic heterocycles. Reagents and conditions (a) Cu(OAc)$_2$, H$_2$O, O$_2$, DMF, 130 °C, 69 %; (b) triphosgene, TMEDA, CH$_2$Cl$_2$, 0 °C to rt, 70 % [2 steps (a, b)]; (c) thiophosgene, Et$_3$N, CH$_2$Cl$_2$, 0 °C to rt, >99 %; (d) BrCN, CH$_2$Cl$_2$, rt, 34 %; (e) Cu(OAc)$_2$, BocNH$_2$, O$_2$, DMF, 130 °C, 53 %; (f) Lawesson's reagent, xylene, reflux, 19 %; (g) Cu(OAc)$_2$, CS$_2$, O$_2$, 1,4-dioxane, 130 °C, 11 %; (h) NaOH, MeOH, H$_2$O, reflux; (i) BrCN, EtOH, reflux, 61 % [2 steps (h, i)]; (j) triphosgene, Et$_3$N, CH$_2$Cl$_2$, 0 °C to rt, 65 % [2 steps (h, j)]

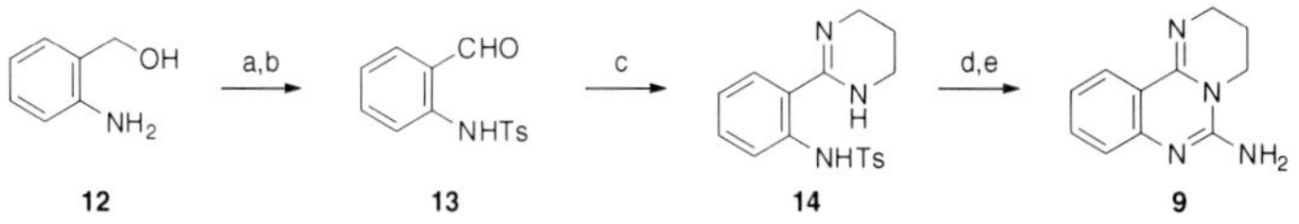

Scheme 3.2 Synthesis of 2-aminoquinazoline derivative **9**. Reagents and conditions (a) *p*-TsCl, pyridine, CHCl$_3$, rt; (b) PCC, silica gel, CH$_2$Cl$_2$, rt, 80 % [2 steps (a, b)]; (c) 1,3-propanediamine, I$_2$, K$_2$CO$_3$, *t*-BuOH, 70 °C, 98 %; (d) conc. H$_2$SO$_4$, 100 °C, then NaOH, H$_2$O; (e) BrCN, EtOH, reflux, 66 % [2 steps (d, e)]

Lawesson's reagent led to formation of the thiocarbonyl derivative **8**. Since no hydrolysis of the carbonyl or thiocarbonyl group of compound **7** or **8** for construction of the 2-aminoquinazoline structure in **9** occurred, an alternative approach starting from 2-aminobenzyl alcohol **12** was used for the synthesis the 2-aminoquinazoline derivative **9** (Scheme 3.2). After protection and PCC oxidation of **12**, oxidative amidation [3] provided 2-(*p*-tosylamino)phenyltetrahydropyrimidine (**14**). Deprotection followed by BrCN-mediated cyclization of **14** provided the expected 2-aminoquinazoline derivative **9**.

To synthesize pyrimido[1,2-*c*][1,3]benzothiazine derivatives **1** and **11** (Scheme 3.1), compound **2** was exposed to CS$_2$ in the presence of Cu(OAc)$_2$ to directly afford a pyrimido[1,2-*c*][1,3]benzo-thiazine-6-thione scaffold **10**. Hydrolysis of the thiocarbonyl group in **10** followed by treatment with BrCN or triphosgene provided 6-imino or 6-oxo derivatives (**1** or **11**), respectively.

Pyrimido[1,2-*c*][1,3]thiazin-6-imine derivatives **25–27** with a series of fused benzene and heterocycles were prepared by consecutive heterocumulene addition and S$_N$Ar reactions (Scheme 3.3). These reactions provide easy access to the construction of the 1,3-thiazin-2-imine derivatives more efficiently (Chap. 2.2) than the diversity-oriented C–H functionalization approach (Chap. 2.1). The oxidative amidation of aromatic aldehydes **15–17** with an accessory functional group afforded the corresponding 2-phenyltetrahydropyrimidine derivatives **18–20**. The pyrimido[1,2-*c*][1,3]thiazine-6-thione scaffold **21** was obtained by additions of **18f,g,i** or **20s,t,u** to CS$_2$ followed by S$_N$Ar-type C–S bond formation. The desired 6-imino derivatives **25f,g,i** and **27s,t,u** were obtained via hydrolysis of the thiocarbonyl group of **21** followed by BrCN treatment. Alternatively, reactions of other 2-phenyltetrahydropyrimidines **18–20** with *tert*-butyl isothiocyanate afforded *N*-(*t*-Bu)-protected thiazinimine derivatives **22–24**, which were treated with TFA to provide the expected products **25–27**.

The intermediates **22e**, **22k**, and **23k** were subjected to further manipulations to obtain the functionalized derivatives (Scheme 3.4). The nitro group of **22e** was reduced by hydrogenation to form the 9-amino derivative **28**. Alkylation of **28** afforded the 9-(*N*-methylamino) derivative **22b** (eq 1). The 9-acetamide derivative **22c** was obtained by treatment of **28** with acetic anhydride (eq 2). Sandmeyer reaction of **28** gave the 9-azide derivative **22p** (eq 3). Me$_2$N- and MeO-substituted derivatives (**22a**, **23a**, and **23f**) were obtained by Me$_2$NH-mediated *N*-arylation [4, 5] of the 9-bromo **22k** and 10-bromo derivatives **23k**, and NaOMe-mediated

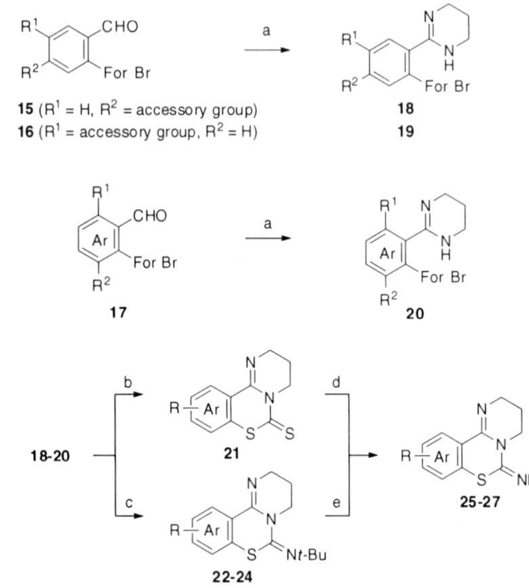

Scheme 3.3 Synthesis of pyrimido[1,2-*c*][1,3]thiazin-6-imine derivatives fused with substituted benzene and heterocycles (**25–27**). Reagents and conditions (a) 1,3-propanediamine, I_2, K_2CO_3, *t*-BuOH, 70 °C, 58–91 %; (b) NaH, CS_2, DMF, 80 °C, 67 to >99 %; (c) NaH or *t*-BuOK, *t*-BuNCS, DMF or DMAc, rt–80 °C, 28–95 %; (d) (i) NaOH, MeOH, H_2O, reflux, (ii) BrCN, EtOH, reflux, 32–68 %; (e) TFA, MS4Å, $CHCl_3$, reflux, 63–92 %

Ullmann coupling [6] of **23k**, respectively (eq 4 and 7). The 9-acetyl derivative **22d** was obtained by Heck reaction [7] of **22k** with 2-hydroxyethyl vinyl ether (eq 5). Other derivatives with a variety of functional groups (**22**, **23**, **29**, and **30**) were synthesized by Suzuki–Miyaura coupling reactions [8, 9] of **22k** and **23k** with boronic acids or their pinacol esters (eq 6 and 7). Final deprotection of the *tert*-butyl group in **22**, **23**, **29**, and **30** afforded the 9- or 10-substituted pyrimido[1,2-*c*][1,3]benzothiazine derivatives **25**, **26**, **31**, and **32**, respectively.

Benzo[*e*][1,3]thiazine derivatives with various ring-sized and/or modified cyclic amidine moieties **36** were synthesized using standard synthetic methods (Scheme 3.5). Oxidative addition using a number of diamines **33** proceeded efficiently to form five- or six-membered rings (**34a–d**). The same reaction for the seven-membered amidine (**34e**) was incomplete, but purification of the Boc-protected amidine **37** followed by subsequent deprotection of the Boc group gave the pure seven-membered amidine **34e**. The resulting amidines were converted to cyclic-amidine-fused benzo[*e*][1,3]thiazin-2-imines **35** via *tert*-butyl isothiocyanate addition and an S_NAr reaction. TFA-mediated deprotection gave the expected derivatives **36**.

The synthesis of the spiropyrimidine-fused derivatives started with the dialkylation of malononitrile with dihaloalkanes (**38**, **39**, or **41**, Scheme 3.6). BH_3-mediated reduction of the alkylated malononitriles (**42–44**) followed by oxidative amidination with 4-bromo-2-fluorobenzaldehyde gave the 2-phenyl-1,4,5,6-tetrahydropyrimidine derivatives (**45–47**). Subsequent exposure of compounds **45–47** to *tert*-butylisothiocyanate provided the tetracyclic compounds **48**, **50**, and **52a**. Deprotection of the *tert*-butyl groups in compounds **48**, **50**, and **52a** afforded the desired spiropyrimidine-fused benzothiazinimine derivatives (**49**, **51**, and **53a**).

The substitution of the *p*-methoxybenzyl (PMB) group in compound **53a** was also attempted (Scheme 3.6). The treatment of compound **52a** with methyl chloroformate or acetyl chloride directly provided derivatives **52b** and **52c**, respectively. A two-step procedure, including the removal of the PMB group by treatment with 1-chloroethyl chloroformate followed by modification with mesyl chloride (MsCl) or trimethylsilyl isocyanate (TMSNCO) was used for the synthesis of the derivatives **52d** and **52e**, respectively, because the reaction of compound **52a** with MsCl and TMSNCO failed. Deprotection of the *tert*-butyl group in **52b–e** afforded the respective *N*-substituted derivatives **53b–e**.

SARs of the central heterocyclic core in pyrimido[1,2-*c*][1,3]benzothiazines were carried out. Initially, the structural requirements of the 1,3-thiazin-2-imine core substructure in **1** (PD 404182) for anti-HIV activity were investigated (Table 3.1). The antiviral activities against the HIV-1$_{IIIB}$ strain were evaluated using the MAGI assay [10]. Substitution of the imino group in **1** with a carbonyl group (**11**) resulted in a significant decrease in anti-HIV activity (EC_{50} = 8.94 μM). Pyrimido[1,2-*c*][1,3]benzoxazines (**4–6**), pyrimido[1,2-*c*] quinazolines (**7–9**), and pyrimido[1,2-*c*] [1,3]benzothiazine-6-thione (**10**), in which the 1-sulfur and/or 2-imino groups in **1** were modified, showed no activity. These results suggested that both the 1-sulfur atom and the 2-imino group are indispensable functional groups for the inhibitory activity against HIV infection, and may be involved in potential interactions with the target molecules.

A series of derivatives with modification of the benzene substructure in the pyrimido[1,2-*c*][1,3]benzothiazine were evaluated for anti-HIV activity (Table 3.2). The addition of positively charged *N*,*N*-dimethylamino (**25a**) and *N*-methylamino groups (**25b**) at the 9-position significantly decreased the anti-HIV activity. The 9-acetamide group (**25c**), which has hydrogen bond donor/acceptor abilities, also attenuated the bioactivity. The acetyl (**25d**) and nitro (**25e**) groups, with hydrogen acceptor properties, induced slight decreases in the anti-HIV activity. In contrast, derivatives with less-polarized substituents (**25f–o** and **25q**) at this position generally reproduced the potent anti-HIV activity of **1**. In terms of the electron-donating or -withdrawing properties of the substituent groups on the benzene substructure, good correlations were not observed. For example, the electron-donating methoxy (**25f**), methyl (**25g**), and *n*-butyl groups (**25h**), and the electron-withdrawing fluoro (**25i**) and trifluoromethyl groups (**25j**) exhibited similar anti-HIV activities (EC_{50} = 0.44–0.57 μM), indicating that the antiviral activity is independent of the electronic state of the 1,3-benzothiazin-2-imine core in forming potential π-stacking interaction(s) with the target molecules. Among

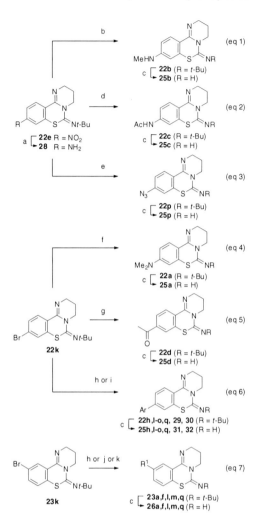

Scheme 3.4 Synthesis of 9- or 10-substituted pyrimido[1,2-*c*][1,3]benzothiazin-6-imine deriv-atives. Reagents and conditions (a) H_2, 10 % Pd/C, EtOH, rt, 88 %; (b) NaOMe, $(CH_2O)_n$, MeOH, reflux, then $NaBH_4$, 91 %; (c) TFA, MS4Å, $CHCl_3$, reflux, 37–95 %; (d) Ac_2O, DMAP, Et_3N, CH_2Cl_2, rt, >99 %; (e) $NaNO_2$, AcOH, H_2O, 0 °C, then NaN_3, 70 %; (f) $Pd(OAc)_2$, *t*-Bu_3P, $NHMe_2$, THF, KO*t*-Bu, toluene, reflux, >99 %; (g) 2-hydroxyethylvinylether, $Pd(OAc)_2$, 1,3-bis(diphenylphosphino)propane, K_2CO_3, H_2O, 90 °C, 13 % [2 steps (g, c)]; (h) R-B(OH)$_2$ or R-Bpin, $Pd(PPh_3)_4$, $PdCl_2$(dppf) · CH_2Cl_2, K_2CO_3, toluene or 1,4-dioxane, EtOH, H_2O, reflux, 62–96 %; (i) *n*-BuB(OH)$_2$, Pd_2(dba)$_3$, P(*t*-Bu)$_3$, $CsCO_3$, 1,4-dioxane, reflux, 6 % (for **22h**); (j) Pd(P*t*-Bu$_3$)$_2$, $NHMe_2$, THF, KO*t*-Bu, toluene, 170 °C, 67 % (for **23a**); (k) CuBr, NaOMe, MeOH, DMF, reflux, 40 % (for **23f**)

the hydrophobic substituents at this position, bromo (**25k**), phenyl (**25l**), vinyl (**25m**), styryl (**25n**), and pentenyl groups (**25o**) induced inhibitory activity two or three times greater than that of **1** (EC_{50} = 0.18–0.25 µM). Modification with photoreactive azido (**25p**) and benzoylphenyl groups (25q) maintained the inhibitory activity; these could be used as probe molecules to identify the target molecule(s) of **1** [11–13].

Similar SARs were observed for modification at the 10-position of pyrimido[1,2-*c*] [1,3]benzothiazine. Addition of positively charged *N*,*N*-dimethylamino (**26a**) and polarized nitro groups (**26e**) reduced the anti-HIV activity (EC_{50} = 2.12 and 3.00 µM, respectively). Hydrophobic groups including methoxy (**26f**), methyl (**26g**), bromo (**26k**), phenyl (**26l**), vinyl (**26m**), and 4-benzoylphenyl (**26q**) had favorable effects on the bioactivity (EC_{50} = 0.24–0.67 µM), suggesting potential hydrophobic interactions of these additional functional groups with the target molecule(s).

Further miscellaneous modifications of benzothiazine substructure were also investigated (Table 3.2). The naphtho[2,3-*e*][1,3]thiazine derivative **27r**, with a 9,10-fused benzene, exhibited anti-HIV activity equipotent to that of the parent **1** (EC_{50} = 0.56 µM). A 6-fold decrease in the anti-HIV activity of the pyridine-fused pyrido[3,2-*e*][1,3]thiazine derivative (**27s**) was observed (EC_{50} = 2.55 µM). In addition, introduction of 8-bromo (**27k**) and 8,9-fused benzene (**27t**, naphtho[2,1-*e*][1,3]thiazine) substituents on benzothiazine resulted in a loss of activity, suggesting that modification at the 8-position was inappropriate for favorable interactions with the target molecule(s). The 11-fluoro derivative **27i** and thiophene-fused **27u**, the latter of which has 5-6-6 framework (thieno[2,3-*e*][1,3]thiazine), exhibited four times lower and no inhibitory potencies, respectively.

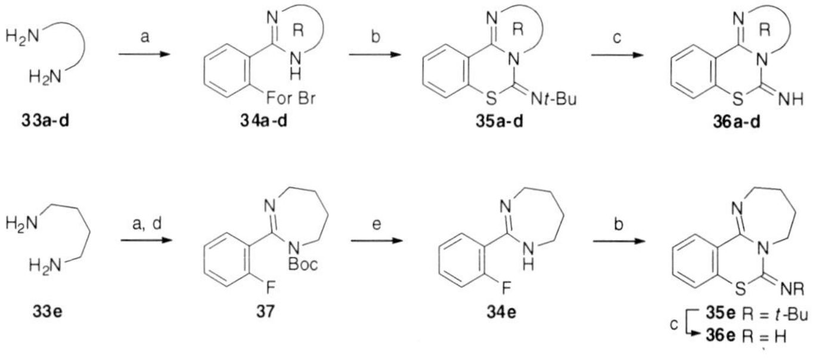

Scheme 3.5 Synthesis of benzo[*e*][1,3]thiazine derivatives with fused cyclic amidines. Reagents and conditions (a) 2-fluorobenzaldehyde or 2-bromobenzaldehyde, I₂, K₂CO₃, *t*-BuOH, 70 °C, 68–79 %; (b) NaH, *t*-BuNCS, DMF, rt −80 °C, 18–50 %; (c) TFA, MS4Å, CHCl₃, reflux, 16–86 %; (d) Boc₂O, Et₃N, DMAP, CH₂Cl₂, rt, 37 % [2 steps (a, d)]; (e) TFA, CHCl₃, reflux, 80 %

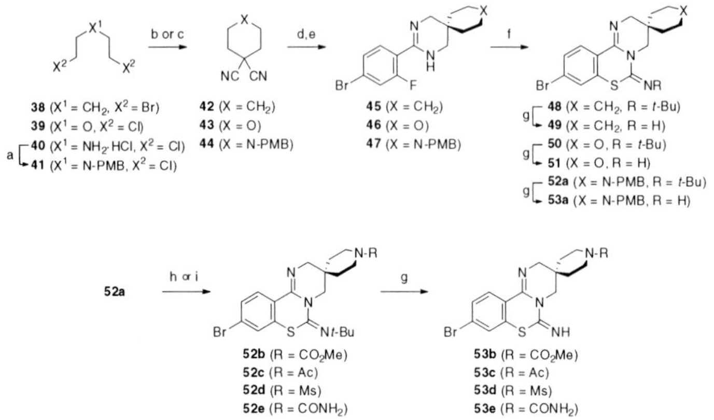

Scheme 3.6 Synthesis of spiropyrimidine-fused benzothiazinimine derivatives. Reagents and conditions (a) (i) 4-methoxybenzoyl chloride, Et$_3$N, CH$_2$Cl$_2$, rt; (ii) LiAlH$_4$, Et$_2$O, rt, 75 % (2 steps); (b) malononitrile, DBU, DMF, 50 °C, 8–60 % (for **42** and **43**); (c) malononitrile, K$_2$CO$_3$, DMF, 65 °C, 85 % (for **44**); (d) BH$_3$, THF, 0 °C to rt; (e) 4-bromo-2-fluorobenzalde-hyde, I$_2$, K$_2$CO$_3$, t-BuOH, 70 °C, 11–62 % [2 steps (d,e)]; (f) NaH, t-BuNCS DMF, rt −80 °C, 78–94 %; (g) TFA, MS4Å, CHCl$_3$, reflux, 65-94 %. (h) ClCO$_2$Me or AcCl, CH$_2$Cl$_2$, 0 °C, 81–96 % (for **52b** or **52c**); (i) (i) 1-chloroethyl chloroformate, Et$_3$N, CH$_2$Cl$_2$, 0 °C, then MeOH, reflux, (ii) MsCl or TMSNCO, (Et$_3$N), CH$_2$Cl$_2$, rt, 29–82 % (2 steps, for **52d** or **52e**)

Table 3.1 SARs for 1,3-thiazin-2-imine core

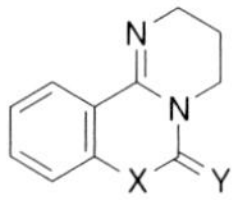

Compound	X	Y	EC$_{50}$ (µM)[a]
1	S	NH	0.44 ± 0.08
4	O	O	>10
5	O	S	>10
6	O	NH	>10
7	NH	O	>10
8	NH	S	>10
9	NH	NH	>10
10	S	S	>10
11	S	O	8.94 ± 1.07

[a] EC$_{50}$ values represent the concentration of compound required to inhibit the HIV-1 infection by 50 % and were obtained from three independent experiments

On the basis of the above SAR data for the benzene substructure in **1** (PD 404182), the author expected that introduction of a hydrophobic group at the pyrimido[1,2-*c*][1,3]benzothiazine 9-position would be the most promising. The next optimization to obtain more potent derivatives was therefore focused on

Table 3.2 SARs for benzene part

Compound		EC_{50} $(\mu M)^a$	Compound		EC_{50} $(\mu M)^a$
			27r		0.56 ± 0.13
1	R = H	0.44 ± 0.08	**27s**		2.55 ± 0.26
25a	R = NMe$_2$	4.74 ± 1.07			
25b	R = NHMe	>10			
25c	R = NHAc	>10			
25d	R = COMe	1.44 ± 0.33			
25e	R = NO$_2$	1.13 ± 0.18			
25f	R = OMe	0.57 ± 0.09	**27k**		>10
25g	R = Me	0.49 ± 0.10			
25h	R = n-butyl	0.44 ± 0.09			
25i	R = F	0.50 ± 0.07			
25j	R = CF$_3$	0.53 ± 0.12			
25k	R = Br	0.25 ± 0.09			
25l	R = Ph	0.24 ± 0.04			
25m	R = vinyl	0.18 ± 0.05	**27t**		>10
25n	R = styryl	0.25 ± 0.05			
25o	R = pentenyl	0.24 ± 0.11			
25p	R = N$_3$	0.43 ± 0.06			
25q	R = C$_6$H$_4$(4-Bz)	0.53 ± 0.12			
			27i		1.68 ± 0.19
26a	R = NMe$_2$	2.12 ± 0.26	**27u**		>10
26e	R = NO$_2$	3.00 ± 0.59			
26f	R = OMe	0.53 ± 0.04			
26g	R = Me	0.38 ± 0.04			
26k	R = Br	0.24 ± 0.05			
26l	R = Ph	0.24 ± 0.05			
26m	R = vinyl	0.40 ± 0.09			
26q	R = C$_6$H$_4$(4-Bz)	0.67 ± 0.16			

[a] EC_{50} values represent the concentration of compound required to inhibit the HIV-1 infection by 50 % and were obtained from three independent experiments

modification of the benzothiazine scaffold at position 9 with an additional aryl group (Tables 3.3, 3.4).

The author initially examined substituent effects at the *para*-position on the 9-phenyl group of compound **25l**. Modification with methoxycarbonyl (**31a**), cyano (**31b**), nitro (**31c**), and trifluoromethyl (**31d**) groups slightly reduced the anti-HIV activity (EC_{50} = 0.44–0.81 μM), whereas a significant decrease in the anti-HIV activity was observed for a carbamoyl group (**31e**), with hydrogen bond donor/acceptor properties (EC_{50} = 8.71 μM). The hydrophobic methoxy (**31f**,

Table 3.3 SARs for biphenyl-type derivatives

Compound	Ar	EC$_{50}$ (μM)^a	Compound	Ar	EC$_{50}$ (μM)^a
	(para-R-phenyl)			(ortho-R-phenyl)	
25l	R = H	0.24 ± 0.04	31s	R = OMe	0.41 ± 0.10
31a	R = CO$_2$Me	0.81 ± 0.29	31t	R = Ph	0.32 ± 0.12
31b	R = CN	0.44 ± 0.10	31u	(3,4-dimethoxyphenyl)	0.27 ± 0.04
31c	R = NO$_2$	0.46 ± 0.06			
31d	R = CF$_3$	0.55 ± 0.16			
31e	R = CONH$_2$	8.71 ± 0.82			
31f	R = OMe	0.24 ± 0.04	31v	(3,4,5-trimethoxyphenyl)	0.25 ± 0.03
31g	R = SMe	0.20 ± 0.06			
31h	R = OCF$_3$	0.38 ± 0.06			
	(meta-R-phenyl)		31w	(chloro-methoxyphenyl)	0.32 ± 0.04
31i	R = CO$_2$Me	0.39 ± 0.09			
31j	R = CN	1.17 ± 0.27			
31k	R = NO$_2$	1.26 ± 0.13			
31l	R = CH(OH)CH$_3$	1.19 ± 0.19		(chloro-methoxyphenyl)	0.48 ± 0.06
31m	R = NHAc	>10			
31n	R = NHMs	>10			
31o	R = OH	2.62 ± 0.26			
31p	R = OMe	0.15 ± 0.05			
31q	R = Oi-Pr	0.32 ± 0.10			
31r	R = Ph	1.35 ± 0.26			

^{a} EC$_{50}$ values represent the concentration of compound required to inhibit the HIV-1 infection by 50 % and were obtained from three independent experiments

EC$_{50}$ = 0.24 μM), methylthio (**31g**, EC$_{50}$ = 0.20 μM), and trifluoromethoxy (**31h**, EC$_{50}$ = 0.38 μM) groups had favorable effects on the anti-HIV activity.

Similar effects as a result of modification at the *meta*-position of the 9-phenyl group were observed. Addition of electron-withdrawing methoxycarbonyl (**31i**), cyano (**31j**), and nitro (**31k**) (EC$_{50}$ = 0.39–1.26 μM) groups resulted in slight decreases in anti-HIV activity. Hydrophilic (1-hydroxy)ethyl (**31l**, EC$_{50}$ = 1.19 μM), acetylamino (**31m**), mesylamino (**31n**), and hydroxyl (**31o**, EC$_{50}$ = 2.62 μM) groups induced reduction or loss of anti-HIV activity. In contrast, a methoxy group (**31p**) improved the inhibitory activity (EC$_{50}$ = 0.15 μM). The more hydrophobic isopropoxy group (**31q**) maintained the anti-HIV activity of 25l (EC$_{50}$ = 0.32 μM), whereas a phenyl group (**31r**) decreased the inhibitory activity (EC$_{50}$ = 1.35 μM).

Table 3.4 SARs for biaryl-type derivatives

Compound	Ar	EC$_{50}$ (μM)[a]	Compound	Ar	EC$_{50}$ (μM)[a]
32a		0.20 ± 0.06	**32h**		0.45 ± 0.07
32b		0.39 ± 0.12	**32i**		0.54 ± 0.04
32c		0.15 ± 0.03	**32j**		0.26 ± 0.02
32d		0.26 ± 0.07	**32 k**		0.20 ± 0.03
32e		0.25 ± 0.04	**32l**		0.22 ± 0.07
32f		$>1.00^{b}$	**32 m**		0.26 ± 0.06
			32n		0.42 ± 0.08
32g		$>1.00^{b}$	**32o**		5.12 ± 1.28

[a] EC$_{50}$ values represent the concentration of compound required to inhibit the HIV-1 infection by 50 % and were obtained from three independent experiments
[b] Cytotoxicity was observed at 10 μM

Similar anti-HIV activities of the *ortho*-methoxy (**31s**) and *ortho*-phenyl group (**31t**) to that of **25l** were exhibited (EC$_{50}$ = 0.41 and 0.32 μM, respectively), suggesting that the twisted conformations of these 9-phenyl PD 404182 derivatives might not prevent the interaction with the target molecule(s).

In order to develop more potent anti-HIV agents, the author subsequently attempted bis and tris modifications of the 9-phenyl group in **25l**. Modification with 9-(3,4-dimethoxy)phenyl (**31u**, EC$_{50}$ = 0.27 μM) or 9-(3,4,5-trimethoxy)phenyl (**31v**, EC$_{50}$ = 0.25 μM) groups of the pyrimido[1,2-c][1,3]benzothiazine scaffold did not alter the bioactivity. Cl-modified derivatives **31w** and **31x** exhibited similar potencies (EC$_{50}$ = 0.32 and 0.48 μM, respectively).

Since the 9-(2-naphthyl)-modified analog (**32a**) exhibited slightly more potent anti-HIV activity (EC$_{50}$ = 0.20 μM) compared with that of the 1-naphthyl congener (**32b**, EC$_{50}$ = 0.39 μM, Table 3.4), the author further investigated modifications with a variety of 3,4-fused phenyl groups. Compound **32c** with a

1,3-dioxolane-fused phenyl group displayed activity twice as potent as that of compound **25l** (EC$_{50}$ = 0.15 µM), whereas the 1,4-dioxane-fused derivative **32d** and quinolin-6-yl derivative **32e** exhibited less favorable effects (EC$_{50}$ = 0.26 and 0.25 µM, respectively). Introduction of an indolyl group (**32f** and **32g**) resulted in no anti-HIV activity and unexpected cytotoxicity.

Substitutions of the 9-phenyl group by various heterocyclic substructures were also investigated. Six-membered heterocycles such as pyridine (**32h** and **32i**) slightly reduced the anti-HIV activity (EC$_{50}$ = 0.45 and 0.54 µM, respectively). The five-membered furan (**32j**), benzofuran (**32k**), thiophene (**32l**), benzothiophene (**32m**), and pyrazole (**32n**) derivatives maintained the original activity of **25l** (EC$_{50}$ = 0.20–0.42 µM). Notably, reduced anti-HIV activity was observed for the basic imidazole derivative (**32o**, EC$_{50}$ = 5.12 µM).

A SAR study of the top-right cyclic amidine substructure was carried out. The five-membered dihydroimidazole derivative **36a** had no anti-HIV activity (Table 3.5), suggesting that the five-membered ring may impair the critical interactions with the target molecule(s) via its small-sized ring strain or indirect effects on the thiazinimine core with a possibly altered conformation. Similarly, compound **36b** with the phenyl-fused dihydropyrimidine substructure showed lower inhibitory activity (EC$_{50}$ = 3.78 µM). Appending one or two methyl groups on the six-membered pyrimidine (**36c** and **36d**) induced 1.5- to 2-fold higher inhibitory potencies (EC$_{50}$ = 0.35 and 0.24 µM, respectively) compared with that of the parent compound **1**. In addition, compound **36e** with a seven-membered tetrahydro-1,3-diazepine substructure exhibited similar anti-HIV activity to that of **1** (EC$_{50}$ = 0.31 µM).

Above optimization studies indicated that the introduction of a hydrophobic group on the cyclic amidine substructures effectively improved the antiviral activity (compound **36c–e**) by generating a potentially favorable interaction(s) with the target molecule(s). Therefore, anti-HIV activities of several spiropyrimidine fused derivatives were evaluated (Table 3.6).[1] Cyclohexane (**49**) and N-methoxycarbonylpiperidine (**53b**) derivatives exhibited the similar levels of anti-HIV activity (EC$_{50}$ = 0.25 and 0.44 µM, respectively) to that of the dimethyl derivative **36d** (EC$_{50}$ = 0.24 µM). In contrast, the tetrahydropyran (**51**) and N-(p-methoxybenzyl)piperidine (**53a**) derivatives exerted inhibitory activities that were 5–7-fold lower (EC$_{50}$ = 1.73 and 1.45 µM, respectively) than that of the parent dimethyl derivative **36d**. The N-acetyl- (**53c**), N-methanesulfonyl- (**53d**), and N-carbamoyl- (**53e**) piperidine derivatives also provided reduced levels of antiviral activity (EC$_{50}$ = 1.81 to >10 µM). With this in mind, the N-alkoxycarbonyl piperidine group was identified as a linkage for the introduction of additional functional group(s) to PD 404182 with potent anti-HIV activity (**53b**).

To investigate the mechanism of action of PD 404182 derivatives, a time of drug addition study was carried out (Fig. 3.3). In this experiment, the anti-HIV

---

[1] Because a 9-brominated derivative **25k** exhibited more potent anti-HIV activity than compound 1 in the SAR study, the author employed compound **25k** as a lead.

Table 3.5 SARs for cyclic amidine core

Compound		EC$_{50}$ (µM)[a]	Compound		EC$_{50}$ (µM)[a]
1		0.44 ± 0.08	36c		0.35 ± 0.09
36a		>10	36d		0.24 ± 0.04
36b		3.78 ± 1.39	36e		0.31 ± 0.06

[a] EC$_{50}$ values represent the concentration of compound required to inhibit the HIV-1 infection by 50 % and were obtained from three independent experiments

Table 3.6 SARs for spiropyrimidine-fused derivatives

Compound	X	EC$_{50}$ (µM)[a]
49	CH$_2$	0.25 ± 0.01
51	O	1.73 ± 0.35
53a	N-PMB	1.45 ± 0.05
53b	N-CO$_2$Me	0.44 ± 0.02
53c	N–Ac	2.74 ± 0.15
53d	N-Ms	1.81 ± 0.43
53e	N-CONH$_2$	>10

[a] EC$_{50}$ values represent the concentration of compound required to inhibit the HIV-1 infection by 50 % and were obtained from three independent experiments

activity profiles of **1** and its derivatives **32c** were compared with those of well-known anti-HIV agents such as an adsorption inhibitor (DS 5000) [14], fusion inhibitor (enfuvirtide) [15–17], NRTI (AZT) [18], NNRTI (nevirapine) [19], and integrase inhibitor (raltegravir) [20]. After inoculation of HeLa-CD4/CCR5-LTR/ β-gal cells with HIV-1$_{IIIB}$, each anti-HIV-1 drug was added at a 90 % inhibitory effect concentration at the indicated time points. The inhibitory effects on the infection were determined by counting the blue cells 48 h later. This investigation revealed that compound **1** (PD 404182) had an inhibitory profile in the early stage

Fig. 3.3 Time of drug
addition profiles in the
presence of Anti-HIV agents
for infection of a HIV-1$_{IIIB}$
strain of HeLa-CD4/CCR5-
LTR/β-gal cells

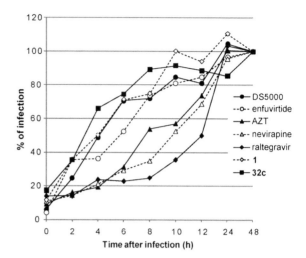

of viral infection similar to those of DS 5000 and enfuvirtide (Fig. 3.3). Identical
profiles were observed for **1** and the most potent derivative **32c**, indicating that the
bioactivity profile is independent of the appended functional group(s).

To gain additional insights into the mechanism of action of PD 404182
derivatives, the antiviral activities against other HIV subtypes were evaluated
(Table 3.7). Compound **1** was effective against not only HIV-1$_{IIIB}$ but also other
two HIV-1 strains (HIV-1$_{NL4-3}$ and HIV-1$_{BaL}$) with similar potency. Both HIV-
1$_{IIIB}$ and HIV-1$_{NL4-3}$ strains utilize CXCR4 as a coreceptor for entry, while HIV-
1$_{BaL}$ strain does CCR5, indicating that chemokine receptors CXCR4 and CCR5 are
not the molecular targets of PD 404182 derivatives. The similar level of antiviral
activity of **1** against HIV-2 (HIV-2$_{EHO}$ and HIV-2$_{ROD}$), which is mainly distrib-
uted in West Africa, was observed. Highly potent inhibitory activities of derivatives **32c** and **36d**[2] against these HIV strains were observed, as in the case of the
SAR study of the HIV-1$_{IIIB}$ strain discussed above. It has been well-known that
NNRTIs are not effective against HIV-2, highlighting that PD 404182 derivatives
do not act as NNRTIs. Although PD 404182 derivatives and enfuvirtide showed
similar anti-HIV-1 profile in the time of drug addition assay, HIV-2$_{EHO}$ and HIV-
2$_{ROD}$ infection were affected by PD 404182 derivatives, in contrast with the less
effective enfuvirtide [21], suggesting that PD 404182 derivatives may not be
directed at the HIV gp41 envelope protein. Recent reports have suggested that the
antiviral activities of compound **1** against HIV, HCV, and pseudotype lentiviruses
were derived from disruption of the structural integrities of virions [2]. Although
the mechanism of action of PD 404182 derivatives is not fully understood at this

---

[2] The cytotoxicity of compounds **1**, **32c** and **36d** was not observed at 10 μM in the MAGI assay.
Further toxicity studies such as hemolytic activity or renal/liver accumulation may be needed to
take a drug for long periods of time.

Table 3.7 Anti-HIV activity of compounds **1**, **32c**, and **36d** against other HIV strains

Strains	EC_{50} $(\mu M)^a$		
	1	**32c**	**36d**
HIV-1$_{NL4-3}$	0.38 ± 0.06	0.25 ± 0.03	0.23 ± 0.09
HIV-1$_{BaL}$	0.37 ± 0.06	0.16 ± 0.02	0.13 ± 0.05
HIV-2$_{EHO}$	0.31 ± 0.06	0.17 ± 0.03	0.14 ± 0.02
HIV-2$_{ROD}$	0.30 ± 0.06	0.11 ± 0.03	0.10 ± 0.04

[a] EC_{50} values represent the concentration of compound required to inhibit the HIV-1 infection by 50 % and were obtained from three independent experiments

stage, the unidentified biomolecule(s) in viruses or host cells could be promising molecular targets for this new class of anti-HIV agents.

In conclusion, the author have designed and synthesized PD 404182 derivatives for a novel series of anti-HIV agents. Comprehensive SAR studies demonstrated that the 6-6-6 fused pyrimido[1,2-*c*][1,3]benzothiazine scaffold and the heteroatom arrangement in the thiazinimine moiety are indispensable for the inhibitory activity of **1** (PD 404182) against HIV infection. Optimization studies of the benzene and cyclic amidine rings indicate that the introduction of a hydrophobic group on the benzene ring is more effective in improving the antiviral activity, giving potential favorable interaction(s) with the target molecule(s). The most potent compound, **32c**, had anti-HIV activity three times higher than that of the parent **1**. In addition, PD 404182 derivatives could be promising agents for treatment of HIV-2 infection. The author also revealed, using a time of drug addition experiment, that PD 404182 derivatives prevent the HIV infection process at a fusion or binding process.

3.1 Experimental Section

3.1.1 General Methods

All moisture-sensitive reactions were performed using syringe-septum cap techniques under an Ar atmosphere and all glasswares were dried in an oven at 80 °C for 2 h prior to use. Melting points were measured by a hot stage melting point apparatus (uncorrected). For flash chromatography, Wakogel C-300E (Wako) or aluminum oxide 90 standardized (Merck) was employed. For preparative TLC, TLC silica gel 60 F254 (Merck) or TLC aluminum oxide 60 F254 basic (Merck), or NH$_2$ Silica Gel 60 F254 Plate (Wako) were employed. For analytical HPLC, a COSMOSIL 5C18-ARII column (4.6 × 250 mm, Nacalai Tesque, Inc., Kyoto, Japan) was employed with method A [a linear gradient of CH$_3$CN containing 0.1 % (v/v) TFA] or method B [a linear gradient of CH$_3$CN containing 0.1 % (v/v) NH$_3$] at a flow rate of 1 mL/min on a Shimadzu LC-10ADvp (Shimadzu Corp., Ltd., Kyoto, Japan), and eluting products were detected by UV at 254 nm.

Preparative HPLC was performed using a COSMOSIL 5C18-ARII column (20 × 250 mm, Nacalai Tesque Inc.) with a linear gradient of MeCN containing 0.1 % (v/v) NH_3 at a flow rate of 8 mL/min on Shimadzu LC-6AD (Shimadzu corporation, Ltd). ^{1}H-NMR spectra were recorded using a JEOL AL-400 or a JEOL ECA-500 spectrometer, and chemical shifts are reported in δ (ppm) relative to Me_4Si ($CDCl_3$) or DMSO (DMSO-d_6) as internal standards. ^{13}C-NMR spectra were recorded using a JEOL AL-400 or JEOL ECA-500 spectrometer and refer- enced to the residual solvent signal. ^{19}F–NMR spectra were recorded using a JEOL ECA-500 and referenced to the internal $CFCl_3$ (δ_F 0.00 ppm). ^{1}H-NMR spectra are tabulated as follows: chemical shift, multiplicity (b = broad, s = singlet, d = doublet, t = triplet, q = quartet, m = multiplet), coupling constant(s), and number of protons. Exact mass (HRMS) spectra were recorded on a JMS-HX/HX 110A mass spectrometer. Infrared (IR) spectra were obtained on a JASCO FT/IR- 4100 FT-IR spectrometer with JASCO ATR PRO410-S. The purity of the com- pounds was determined by combustion analysis or HPLC analysis as >95 % unless otherwise stated. Synthesis and characterization data of compounds **2**, **4**, **7**, and **10** are shown in Chap. 2. Synthesis and characterization data of compounds **1**, **18e-g**, **18i**, **19e**, **20s**, **20t**, **21f**, **21g**, **21i**, **21s**, **21t**, **34a** and **35a** are shown in Chap. 2.

3.1.2 Synthesis of 3,4-Dihydro-2H,6H-pyrimido [1,2-c][1,3]benzoxazine-6-thione (5)

2-(2-Hydroxyphenyl)tetrahydropyrimidine (3). DMF (0.83 mL) and water (4.5 μL, 0.25 mmol) were added to a flask 2-phenyl-1,4,5,6-tetrahydropyrimidine **2** (40.1 mg, 0.25 mmol) and $Cu(OAc)_2$ (45.4 mg, 0.25 mmol) under an O_2 atmosphere. After being stirred at 130 °C for 20 min, mixture was concentrated. The residue was purified by flash chromatography over aluminum oxide with $CHCl_3$–MeOH (95:5) to give the title compound **3** as brown solid (30.3 mg, 69 %): IR (neat) cm^{-1}: 3257–3041 (OH), 1613 (C=N); ^{1}H-NMR (500 MHz, DMSO-d_6) δ: 1.84–1.88 (2H, m, CH_2), 3.40 (4H, t, $J = 5.7$ Hz, 2 × CH_2), 6.27–6.30 (1H, m, Ar), 6.47 (1H, d, $J = 8.6$ Hz, Ar), 7.04–7.08 (1H, m, Ar), 7.45 (1H, dd, $J = 8.0, 1.7$ Hz, Ar), 12.09 (1H, br s); ^{13}C-NMR (125 MHz, CD$_3$OD) δ: 20.0, 39.3 (2C), 111.2, 114.5, 124.4, 126.3, 135.1, 161.0, 172.4; MS (FAB) m/z (%): 177 (MH$^+$, 100).

Compound 5. To a suspension of **3** (33.0 mg, 0.19 mmol) and Et$_3$N (0.068 mL, 0.47 mmol) in CH$_2$Cl$_2$ (10.0 mL) was added dropwise a solution of thiophosgene (0.016 mL, 0.21 mmol) in CH$_2$Cl$_2$ (1.0 mL) at 0 °C. After being stirred at rt for 1 h, the mixture was quenched with sat. NaHCO$_3$. The whole was extracted with EtOAc. The extract was washed with sat. NaHCO$_3$, brine, and dried over MgSO$_4$. After concentration, the residue was purified by flash chromatog- raphy over silica gel with n-hexane–EtOAc (3:1) to give the title compound **5** as yellow solid (41.9 mg, >99 %): mp 135–136 °C (from CHCl$_3$–n-hexane); IR (neat) cm^{-1}: 1655 (C=N); ^{1}H-NMR (400 MHz, CDCl$_3$) δ: 2.03–2.08 (2H, m, CH$_2$), 3.68 (2H, t, J = 5.5 Hz, CH$_2$), 4.30 (2H, t, J = 6.1 Hz, CH$_2$), 7.21 (1H, d, J = 8.5 Hz, Ar), 7.25–7.29 (1H, m, Ar), 7.50–7.52 (1H, m, Ar), 8.00 (1H, d, J = 7.8 Hz, Ar); ^{13}C-NMR (100 MHz, CDCl$_3$) δ: 21.1, 44.5, 49.2, 115.9, 116.9, 125.4, 125.7, 133.0, 139.7, 150.9, 180.8; *Anal.* calcd for C$_{11}$H$_{10}$N$_2$OS: C, 60.53; H, 4.62; N, 12.83. Found: C, 60.23; H, 4.72; N, 12.62.

3.1.3 Synthesis of 3,4-Dihydro-2H,6H-pyrimido [1,2-c][1,3]benzoxazin-6-imine (6)

2-(2-Hydroxyphenyl)tetrahydropyrimidine **3** (5.4 mg, 0.03 mmol) was suspended with CH$_2$Cl$_2$ (0.3 mL) and added the solution of BrCN (3.3 mg, 0.06 mmol) in CH$_2$Cl$_2$ (0.3 mL). After being stirred for 1 h at rt, the additional portion of BrCN (3.3 mg, 0.06 mmol) in CH$_2$Cl$_2$ (0.3 mL) was added. After being stirred for 1 h at rt, the mixture was concentrated. The residue was purified by preparative TLC over NH$_2$ silica gel with *n*-hexane–EtOAc (1:1) to give the title compound **6** as colorless solid (2.1 mg, 34 %): mp 104–105 °C (from CHCl$_3$–*n*-hexane); IR (neat) cm^{-1}: 1639 (C=N), 1611 (C=N); ^{1}H-NMR (500 MHz, CDCl$_3$) δ: 1.98–2.03 (2H, m, CH$_2$), 3.64 (2H, t, J = 5.4 Hz, CH$_2$), 3.93 (2H, t, J = 6.0 Hz, CH$_2$), 5.83 (1H, br s, NH), 6.99 (1H, d, J = 8.0 Hz, Ar), 7.15 (1H, t, J = 8.0 Hz, Ar), 7.42 (1H, td, J = 8.0, 1.7 Hz, Ar), 7.99 (1H, dd, J = 8.0, 1.7 Hz, Ar); ^{13}C-NMR (125 MHz, CDCl$_3$) δ: 20.6, 43.4, 44.1, 115.2, 116.2, 123.9, 125.5, 132.3, 142.5, 150.4, 150.7; *Anal.* calcd for C$_{11}$H$_{11}$N$_3$O: C, 65.66; H, 5.51; N, 20.88. Found: C, 65.55; H, 5.40; N, 20.70.

3.1.4 Synthesis of 3,4-Dihydro-2H,6H-pyrimido [1,2-c]quinazolin-6(7H)-thione (8)

Xylene (4.0 mL) was added to a flask containing 3,4-dihydro-2*H*,6*H*-pyrimido[1,2-*c*]quinazolin-6(7*H*)-one **7** (50.3 mg, 0.25 mmol) and Lawesson's reagent (202.2 mg, 0.50 mmol). After being stirred under reflux for 24 h, xylene (2 mL) and additional amount of Lawesson's reagent (101.1 mg, 0.25 mmol) were added. After being stirred under reflux for additional 12 h, the mixture was cooled to rt. The residue was dissolved in CHCl$_3$ and washed with sat. NaHCO$_3$ and brine and dried over MgSO$_4$. After concentration, the residue was purified by flash chromatography over silica gel with *n*-hexane–EtOAc (2:1) to give the title compound **8** as colorless solid (10.4 mg, 19 %): mp 258–259 °C (from CHCl$_3$–*n*-hexane); IR (neat) cm^{-1}: 1618 (C=N); ^{1}H-NMR (500 MHz, DMSO-d_6) δ: 1.85–1.90 (2H, m, CH$_2$), 3.52 (2H, t, J = 5.4 Hz, CH$_2$), 4.19 (2H, t, J = 6.0 Hz, CH$_2$), 7.12–7.19 (2H, m, Ar), 7.46 (1H, t, J = 7.7 Hz, Ar), 7.92 (1H, d, J = 6.9 Hz, Ar), 12.00 (1H, s, NH); ^{13}C-NMR (100 MHz, DMSO-d_6) δ: 20.6, 44.0, 46.7, 114.8, 117.6, 123.8, 125.1, 132.1, 135.9, 142.1, 174.0; HRMS (FAB): m/z calcd for $C_{11}H_{12}N_3S$ [M + H]$^+$ 218.0752; found: 218.0757.

3.1.5 Synthesis of 3,4-Dihydro-2H,6H-pyrimido [1,2-c]quinazolin-6-amine (9)

2-[N-(p-Toluenesulfonyl)amino]benzaldehyde (13). To a solution of 2-amino-benzylalcohol **12** (2.0 g, 16.2 mmol) and pyridine (1.6 mL, 19.4 mmol) in CHCl$_3$ (60 mL) was added a solution of p-TsCl (3.4 g, 18.0 mmol) in CHCl$_3$ (17 mL), and the mixture was stirred at rt for 3 h. After concentration, EtOAc and sat. NH$_4$Cl were added to the residue. The organic phase was separated and dried over MgSO$_4$. After concentration, the resulting solid was added to a suspension of PCC (5.2 g, 24.3 mmol) and silica gel (10.6 g) in CHCl$_3$ (70 mL). After being stirred at rt for 2 h, the mixture was filtered and concentrated. The residue was purified by flash chromatography over silica gel with n-hexane–EtOAc (9:1) to give the title compound **13** as colorless solid (3.6 g, 80 %): mp 134–136 °C (from CHCl$_3$–n-hexane); IR (neat) cm^{-1}: 1672 (C=O), 1492 (NSO$_2$), 1157 (NSO$_2$); ^{1}H-NMR (400 MHz, CDCl$_3$) δ: 2.36 (3H, s, CH$_3$), 7.16 (1H, t, J = 7.6 Hz, Ar), 7.24 (2H, d, J = 8.5 Hz, Ar), 7.49–7.53 (1H, m, Ar), 7.59 (1H, dd, J = 7.6, 1.5 Hz, Ar), 7.69 (1H, d, J = 8.3 Hz, Ar), 7.77 (2H, d, J = 8.5 Hz, Ar), 9.83 (1H, s, CHO), 10.78 (1H, br s, NH); ^{13}C-NMR (100 MHz, CDCl$_3$) δ: 21.5, 117.8, 121.9, 122.9, 127.3 (2C), 129.7 (2C), 135.8, 136.1, 136.5, 140.0, 144.1, 194.9; Anal. calcd for $C_{14}H_{13}NO_3S$: C, 61.07; H, 4.76; N, 5.09. Found: C, 60.97; H, 4.46; N, 5.05.

2-[2-N-(p-Toluenesulfonylamino)phenyl]-1,4,5,6-tetrahydropyrimidine (14). To a solution of **13** (2.75 g, 10 mmol) in t-BuOH (94 mL) was added propylene-diamine (969 mg, 11 mmol). The mixture was stirred at 70 °C for 30 min, and then K$_2$CO$_3$ (4.15 g, 30 mmol) and I$_2$ (3.17 g, 12.5 mmol) were added. After being stirred at same temperature for 3 h, the mixture was quenched with sat. Na$_2$SO$_3$ until the iodine color disappeared. The organic layer was separated and concentrated. The resulting solid was dissolved in H$_2$O. The whole was extracted with CHCl$_3$, and dried over MgSO$_4$. After concentration, the resulting solid was recrystallized from CHCl$_3$–Et$_2$O–n-hexane to give the title compound **14** as pale yellow crystals (3.23 g, 98 %): mp 211–213 °C; IR (neat) cm^{-1}: 1630 (C=N); 1478 (NSO$_2$); 1124 (NSO$_2$); ^{1}H-NMR (400 MHz, CDCl$_3$) δ: 1.77–1.82 (2H, m, CH$_2$), 2.34 (3H, s, CH$_3$), 3.36 (4H, t, J = 5.7 Hz, 2 × CH$_2$), 6.53–6.57 (1H, m, Ar), 7.04–7.08 (1H, m, Ar), 7.16–7.22 (3H, m, Ar), 7.58 (1H, dd, J = 8.2, 1.3 Hz, Ar), 7.76 (2H, d, J = 8.3 Hz, Ar), 10.75 (1H, br s, NH); ^{13}C-NMR (100 MHz, CDCl$_3$) δ: 18.4, 21.3, 38.8 (2C), 112.4, 117.7, 121.2, 126.3 (2C), 126.5, 129.2 (2C), 133.0, 140.9, 142.0, 150.3, 158.9; HRMS (FAB): m/z calcd for $C_{17}H_{20}N_3O_2S$ [M + H]$^+$ 330.1276; found: 330.1273.

Compound 9. To a flask containing **14** (164.7 mg, 0.5 mmol) was added conc. H_2SO_4 (5.0 mL). After being stirred at 100 °C for 30 min, the mixture was cooled to 0 °C, and then pH was adjusted to 12–14 with 2 N NaOH. The whole was extracted with $CHCl_3$, and dried over $MgSO_4$. After concentration, the residue was dissolved in anhydrous EtOH (2 mL). Then, BrCN (105.9 mg, 1.0 mmol) was added to the mixture under an Ar atmosphere. After being stirred under reflux for 2 h, the reaction was quenched with 2N NaOH. The whole was extracted with $CHCl_3$, and dried over $MgSO_4$. After concentration, the residue was purified by flash chromatography over aluminum oxide with EtOAc–MeOH (95:5) to give the title compound **9** as colorless solid (66.0 mg, 66 %): mp 259–260 °C (from $CHCl_3$–n-hexane); IR (neat) cm^{-1}: 1620 (C=N), 1603 (C=N); ^{1}H-NMR (400 MHz, DMSO-d_6) δ: 1.81–1.87 (2H, m, CH_2), 3.44 (2H, t, J = 5.4 Hz, CH_2), 3.70 (2H, t, J = 6.1 Hz, CH_2), 6.49 (2H, br s, NH_2), 6.87–6.95 (2H, m, Ar), 7.27–7.31 (1H, m, Ar), 7.87 (1H, dd, J = 7.9, 1.1 Hz, Ar); ^{13}C-NMR (100 MHz, DMSO-d_6) δ: 20.0, 42.8, 42.9, 118.9, 120.7, 122.7, 124.3, 131.1, 145.6, 146.6, 151.6; HRMS (FAB): m/z calcd for $C_{11}H_{13}N_4$ [M + H]$^+$ 201.1140; found: 201.1138.

3.1.6 Synthesis of 3,4-Dihydro-2H,6H-pyrimido [1,2-c][1,3]benzothiazin-6-one (11)

3,4-Dihydro-2H,6H-pyrimido[1,2-c][1,3]benzothiazine-6-thione **10** (58.6 mg, 0.25 mmol) was suspended into a 0.1 M NaOH in MeOH-H_2O (9:1, 5 mL). After being stirred under reflux for 12 h, the mixture was concentrated. To a stirring solution of the residue and Et_3N (0.029 mL, 2.0 mmol) in CH_2Cl_2 (16.6 mL) was added dropwise a solution of triphosgene (155.8 mg, 0.52 mmol) in CH_2Cl_2 (1.7 mL) at 0 °C. After being stirred at rt for 1 h, the mixture was quenched with sat. $NaHCO_3$. The whole was extracted with $CHCl_3$. The extract was washed with sat. $NaHCO_3$, brine, and dried over $MgSO_4$. After concentration, the residue was purified by flash chromatography over silica gel with n-hexane–EtOAc (9:1) to give the title compound **11** as colorless solid (35.3 mg, 65 %): mp 102–103 °C (from $CHCl_3$–n-hexane); IR (neat) cm^{-1}: 1639 (C=O) 1612 (C=N); ^{1}H-NMR (500 MHz, $CDCl_3$) δ: 1.95-1.99 (2H, m, CH_2), 3.73 (2H, t, J = 5.7 Hz, CH_2), 4.00 (2H, t, J = 6.0 Hz, CH_2), 7.13 (1H, dd, J = 8.0, 1.3 Hz, Ar), 7.27–7.30 (1H, m, Ar), 7.40 (1H, td, J = 8.0, 1.1 Hz, Ar), 8.28 (1H, dd, J = 8.0, 1.1 Hz, Ar); ^{13}C-NMR (125 MHz, $CDCl_3$) δ: 20.8, 42.4, 45.2, 124.4, 125.8, 126.8, 128.9, 129.2, 130.9, 146.1, 162.8; HRMS (FAB): m/z calcd for $C_{11}H_{11}N_2OS$ [M + H]$^+$ 219.0592; found: 219.0592.

3.1.7 Synthesis of 9-(N,N-Dimethylamino)-3,4-dihydro-2H,6H-pyrimido[1,2-c][1,3]benzothiazin-6-imine (25a)

N-(tert-Butyl)-9-(N',N'H,6H-pyrimido[1,2-c][1,3]benzothiazin-6-imine   (22a).
To   a   mixture   of   N-(tert-butyl)-9-bromo-3,4-dihydro-2H,6H-pyrimido[1,2-c][1,3]benzothiazin-6-imine **22k** (88.1 mg, 0.25 mmol) and Pd(OAc)$_2$ (5.6 mg, 0.025 mmol) and KOt-Bu (84.2 mg, 0.75 mmol) in toluene (2.0 mL) were added P(tert-Bu)$_3$ (0.009 mL, 0.038 mmol) and 2 N Me$_2$NH in THF (0.38 mL, 0.75 mmol). After being stirred at reflux for 1 h, the mixture was filtered through a Celite pad and concentrated. The residue was purified by flash chromatography over aluminum oxide with n-hexane–EtOAc (7:3) to give the title compound **22a** as colorless solid (80.9 mg, >99 %): mp 161–162 °C (from CHCl$_3$–n-hexane); IR (neat) cm^{-1}: 1587 (C=N); ^{1}H-NMR (400 MHz, CDCl$_3$) δ: 1.38 (9H, s, 3 × CH$_3$), 1.86-1.92 (2H, m, CH$_2$), 2.97 (6H, s, 2 × CH$_3$), 3.58 (2H, t, J = 5.5 Hz, CH$_2$), 3.85 (2H, t, J = 6.1 Hz, CH$_2$), 6.28 (1H, d, J = 2.7 Hz, Ar), 6.55 (1H, dd, J = 9.0, 2.7 Hz, Ar), 8.04 (1H, d, J = 9.0 Hz, Ar); ^{13}C-NMR (100 MHz, CDCl$_3$) δ: 22.0, 30.0 (3C), 40.0 (2C), 44.9, 45.5, 54.0, 105.5, 110.6, 115.7, 129.7, 130.0, 139.2, 148.0, 151.2; HRMS (FAB): m/z calcd for C$_{17}$H$_{25}$N$_4$S [M + H]$^+$ 317.1800; found: 317.1803.

   Compound 25a. TFA (2.0 mL) was added to a mixture of **22a** (63.3 mg, 0.2 mmol) in small amount of CHCl$_3$ and MS4 Å (300 mg, powder, activated by heating with Bunsen burner). After being stirred under reflux for 1 h, the mixture was concentrated. To a stirring mixture of the residue in CHCl$_3$ was added dropwise Et$_3$N at 0 °C to adjust pH to 8–9. The whole was extracted with EtOAc. The extract was washed with sat. NaHCO$_3$, brine, and dried over MgSO$_4$. After concentration, the residue was purified by flash chromatography over aluminum oxide with n-hexane–EtOAc (7:3) to give the title compound 25a as colorless solid (38.2 mg, 73 %): mp 150–151 °C (from CHCl$_3$–n-hexane); IR (neat) cm^{-1}: 1600 (C=N), 1562 (C=N); ^{1}H-NMR (500 MHz, CDCl$_3$) δ: 1.93–1.98 (2H, m, CH$_2$), 2.98 (6H, s, 2 × CH$_3$), 3.64 (2H, t, J = 5.7 Hz, CH$_2$), 4.00 (2H, t, J = 6.3 Hz, CH$_2$), 6.17 (1H, d, J = 2.3 Hz, Ar), 6.55 (1H, dd, J = 9.2, 2.3 Hz, Ar), 7.01 (1H, br s, NH), 8.05 (1H, d, J = 9.2 Hz, Ar); ^{13}C-NMR (125 MHz, CDCl$_3$) δ: 21.1, 40.0 (2C), 43.8, 44.7, 104.4, 110.7, 114.4, 129.8, 129.9, 146.7, 151.3, 154.2; HRMS (FAB): m/z calcd for C$_{13}$H$_{17}$N$_4$S [M + H]$^+$ 261.1174; found: 261.1173.

3.1.8 Synthesis of 3,4-Dihydro-9-(N-methylamino)-2H,6H-pyrimido[1,2-c][1,3]benzothiazin-6-imine (25b)

9-Amino-N-(tert-butyl)-3,4-dihydro-2H,6H-pyrimido[1,2-c][1,3]benzothiazin-6-imine (28). To a suspension of N-(tert-Butyl)-3,4-dihydro-9-nitro-2H,6H-pyrimido[1,2-c][1,3]benzothiazin-6-imine **22e** (477.0 mg, 1.5 mmol) in EtOH (10 mL) was added 10 % Pd/C (ca. 55 % in water, 400 mg) under a H_2 atmosphere. After being stirred at rt overnight, the mixture was filtered through a Celite pad. After concentration, the resulting solid was recrystallized from $CHCl_3$–n-hexane to give the title compound **28** as colorless crystals (381.1 mg, 88 %): mp 152–155 °C; IR (neat) cm^{-1}: 1589 (C=N); ^{1}H-NMR (400 MHz, $CDCl_3$) δ: 1.37 (9H, s, 3 × CH_3), 1.86-1.92 (2H, m, CH_2), 3.57 (2H, t, J = 5.4 Hz, CH_2), 3.84 (2H, t, J = 6.0 Hz, CH_2), 3.88 (2H, br s, NH_2), 6.33 (1H, d, J = 2.2 Hz, Ar), 6.49 (1H, dd, J = 8.5, 2.2 Hz, Ar), 7.99 (1H, d, J = 8.5 Hz, Ar); ^{13}C-NMR (100 MHz, CD_3OD) δ: 22.7, 30.2 (3C), 45.2, 46.8, 55.2, 109.1, 114.4, 116.4, 130.7, 131.4, 140.5, 151.8, 152.4; HRMS (FAB): m/z calcd for $C_{15}H_{21}N_4S$ [M + H]$^+$ 289.1487; found: 289.1489.

N-(*tert*-Butyl)-3,4-dihydro-9-(N-methylamino)-2H,6H-pyrimido[1,2-c][1,3] benzothiazin-6-imine (22b). To a flask containing **28** (108.5 mg, 0.38 mmol), MeONa (30.6 mg, 0.57 mmol), and paraformaldehyde (34.2 mg, 1.1 mmol) was added unhydrous MeOH (2.5 mL) under an Ar atmosphere, and stirring was continued for 5 h under reflux. Then, $NaBH_4$ (28.8 mg, 0.76 mmol) was added to the mixture and stirring was continued for additional 30 min under reflux. After concentration, the residue was dissolved in EtOAc, and washed with sat. $NaHCO_3$, brine, and dried over $MgSO_4$. After concentration, the residue was purified by flash column chromatography over aluminum oxide with n-hexane–EtOAc (1:1) to give the title compound **22b** as yellow solid (104.9 mg, 91 %): mp 156–158 °C (from $CHCl_3$–n-hexane); IR (neat) cm^{-1}: 1590 (C=N); ^{1}H-NMR (400 MHz, $CDCl_3$) δ: 1.38 (9H, s, 3 × CH_3), 1.87–1.92 (2H, m, CH_2), 2.83 (3H, s, CH_3), 3.57 (2H, t, J = 5.4 Hz, CH_2), 3.85 (2H, t, J = 6.0 Hz, CH_2), 4.04 (1H, br s, NH), 6.20 (1H, d, J = 2.4 Hz, Ar), 6.43 (1H, dd, J = 8.8, 2.4 Hz, Ar), 8.01 (1H, d, J = 8.8 Hz, Ar); ^{13}C-NMR (100 MHz, $CDCl_3$) δ: 21.9, 30.0 (3C), 30.2, 44.7, 45.5, 54.0, 104.9, 111.8, 116.3, 129.7, 130.4, 139.0, 148.3, 150.6; HRMS (FAB): m/z calcd for $C_{16}H_{23}N_4S$ [M + H]$^+$ 303.1643; found: 303.1638.

Compound 25b. Using the general procedure as described for **25a**, compound **22b** (30.7 mg, 0.1 mmol) was allowed to react for 1 h with TFA (1.0 mL) and MS4Å (200 mg). Purification by flash chromatography over aluminum oxide with n-hexane–EtOAc (1:1 to 0:1) gave the title compound **25b** as pale yellow solid (9.0 mg, 37 %): mp 129–131 °C (from $CHCl_3$–n-hexane); IR (neat) cm^{-1}: 1602 (C=N), 1555 (C=N); ^{1}H-NMR (400 MHz, $CDCl_3$) δ: 1.92–1.98 (2H, m, CH_2), 2.84 (3H, d, J = 4.1 Hz, CH_3), 3.64 (2H, t, J = 5.6 Hz, CH_2), 4.00 (2H, t, J = 6.2 Hz, CH_2), 4.03 (1H, br s, NH), 6.11 (1H, d, J = 2.4 Hz, Ar), 6.44 (1H, dd, J = 8.8, 2.4 Hz, Ar), 8.01 (1H, d, J = 8.8 Hz, Ar); ^{13}C-NMR (100 MHz, $CDCl_3$) δ: 21.1, 30.2, 43.9, 44.6, 104.0, 111.9, 129.2, 130.1, 130.3, 146.9, 150.8, 154.1; *Anal.* calcd for $C_{12}H_{14}N_4S$: C, 58.51; H, 5.73; N, 22.74. Found: C, 58.30; H, 5.62; N, 22.45.

3.1.9 Synthesis of 9-(N-Acetylamino)-3,4-dihydro-2H, 6H-pyrimido[1,2-c][1,3]benzothiazin-6-imine (25c)

9-(N-Acetylamino)-N-(tert-butyl)-3,4-dihydro-2H,6H-pyrimido[1,2-c][1,3]benzothiazin-6-imine (22c). To a mixture of 9-amino-N-(tert-butyl)-3,4-dihydro-2H,6H-pyrimido[1,2-c][1,3]benzothiazin-6-imine **28** (100.9 mg, 0.35 mmol), Et_3N (0.015 mL, 1.05 mmol), DMAP (4.3 mg, 0.04 mmol) in CH_2Cl_2 (3.5 mL) was added Ac_2O (0.066 mL, 0.70 mmol) under an Ar atmosphere. After being stirred under reflux for 1 h, sat. $NaHCO_3$ was added to the mixture. The whole was extracted with EtOAc. The extract was washed with brine, and dried over $MgSO_4$. After concentration, the residue was purified by flash chromatography over aluminum oxide with n-hexane–EtOAc (1:1 to 0:1) to give the title compound **22c** as colorless solid (120.1 mg, > 99 %): mp 213–214 °C (from $CHCl_3$–n-hexane); IR (neat) cm^{-1}: 1680 (C=O), 1596 (C=N); ^{1}H-NMR (400 MHz, $CDCl_3$) δ: 1.37 (9H, s, 3 × CH_3), 1.88–1.93 (2H, m, CH_2), 2.15 (3H, s, CH_3), 3.59 (2H, t, J = 5.4 Hz, CH_2), 3.86 (2H, t, J = 6.1 Hz, CH_2), 6.99 (1H, dd, J = 8.7, 2.1 Hz, Ar), 7.74 (1H, d, J = 2.1 Hz, Ar), 7.96 (1H, br s, NH), 8.08 (1H, d, J = 8.7 Hz, Ar); ^{13}C-NMR (100 MHz, $CDCl_3$) δ: 21.8, 24.6, 29.9 (3C), 44.9, 45.4, 54.2, 114.4, 116.8, 123.1, 129.1, 130.4, 138.2, 139.7, 147.8, 168.6; HRMS (FAB): m/z calcd for $C_{17}H_{23}N_4OS$ [M + H]$^+$ 331.1593; found: 331.1590.

Compound 25c. Using the general procedure as described for **28a**, compound **25c** (120.1 mg, 0.36 mmol) was allowed to react for 10 h. Purification by recrystallization from MeOH–$CHCl_3$–Et_2O gave the title compound 25c as pale yellow crystals (64.9 mg, 65 %): mp 214 °C (decomp.); IR (neat) cm^{-1}: 1681 (C=O), 1619 (C=N), 1550 (C=N); ^{1}H-NMR (400 MHz, DMSO-d_6) δ: 1.85-1.91 (2H, m, CH_2), 2.07 (3H, s, CH_3), 3.55 (2H, t, J = 5.5 Hz, CH_2), 3.92 (2H, t, J = 6.0 Hz, CH_2), 7.32 (1H, dd, J = 8.9, 1.8 Hz, Ar), 7.61 (1H, d, J = 1.8 Hz, Ar), 8.10 (1H, d, J = 8.9 Hz, Ar), 9.14 (1H, br s, NH), 10.27 (1H, s, NH); ^{13}C-NMR (100 MHz, DMSO-d_6) δ: 20.3, 24.1, 43.3, 43.6, 112.2, 116.7, 119.3, 129.2, 129.9, 141.7, 146.3, 149.5, 169.0; HRMS (FAB): m/z calcd for $C_{13}H_{15}N_4OS$ [M + H]$^+$ 275.0967; found: 275.0967.

3.1.10 Synthesis of 9-Acetyl-3,4-dihydro-2H, 6H-pyrimido[1,2-c][1,3]benzothiazin-6-imine (25d)

To a mixture of N-(tert-butyl)-9-bromo-3,4-dihydro-2H,6H-pyrimido[1,2-c][1,3]benzothiazin-6-imine **22k** (100 mg, 0.284 mmol), $Pd(OAc)_2$ (6.4 mg, 0.0284 mmol), K_2CO_3 (120 mg, 0.852 mmol) and dppp (23.7 mg, 0.0568 mmol) in H_2O (0.57 mL) was added ethylene glycol monovinyl ether (0.13 mL, 1.42 mmol). After being stirred at reflux for 12 h, the whole was extracted with $CHCl_3$. The extract was washed with brine, and dried over Na_2SO_4. After concentration, TFA (2.84 mL) was added to resulting residue. After being stirred under reflux for 1.5 h, the mixture was added dropwise to Et_3N at 0 °C to adjust pH to 8–9. The whole was extracted with EtOAc. The extract was washed with brine, and dried over Na_2SO_4. After concentration, the residue was purified by preparative TLC over aluminum oxide with n-hexane–EtOAc (7:3) to give the title compound **25d** as pale yellow solid (9.7 mg, 13 %): mp 148.4 °C; IR (neat) cm^{-1}: 1678 (C=O), 1616 (C=N), 1567 (C=N); ^{1}H-NMR (300 MHz, $CDCl_3$) δ: 1.95-2.03 (2H, m, CH_2), 2.60 (3H, s, CH_3), 3.72 (2H, t, $J = 5.7$ Hz, CH_2), 4.03 (2H, t, $J = 6.3$ Hz, CH_2), 7.63 (1H, d, $J = 1.8$ Hz, Ar), 7.74 (1H, dd, $J = 8.3$, 1.7 Hz, Ar), 8.32 (1H, d, $J = 7.8$ Hz, Ar); ^{13}C-NMR (75 MHz, $CDCl_3$) δ: 20.9, 26.7, 43.8, 45.1, 123.6, 125.7, 129.3, 129.6, 130.4, 138.3, 146.0, 152.6, 196.7; *Anal.* calcd for $C_{13}H_{13}N_3OS$: C, 60.21; H, 5.05; N, 16.20. Found: C, 60.16; H, 5.02; N, 15.94.

3.1.11 Synthesis of 3,4-Dihydro-9-nitro-2H, 6H-pyrimido[1,2-c][1,3]benzothiazin-6-imine (25e)

N-(*tert*-Butyl)-3,4-dihydro-9-nitro-2*H*,6*H*-pyrimido[1,2-*c*][1,3]benzothiazin-6-imine (22e). To a mixture of 2-(2-fluoro-4-nitrophenyl)-1,4,5,6-tetrahydropyrimidine **18e** (2.0 g, 9.0 mmol) and NaH (716.8 mg, 17.9 mmol; 60 % oil suspension) in DMF (29.8 mL) was added *tert*-butylisothiocyanate (2.28 mL, 17.9 mmol) under an Ar atmosphere, and the mixture was stirred at -20 °C to rt for 2 days. The whole was extracted with EtOAc, and the extract was washed with sat. $NaHCO_3$, brine, and dried over Na_2SO_4. After concentration, the residue was purified by flash chromatography over aluminum oxide with n-hexane–EtOAc (1:0 to 9:1) to give the title compound **22e** as pale yellow solid (1.77 g, 62 %): mp 152–153 °C (from $CHCl_3$–n-hexane); IR (neat) cm^{-1}: 1591 (NO_2), 1581 (C=N), 1523 (NO_2); ^{1}H-NMR (500 MHz, $CDCl_3$) δ: 1.39 (9H, s, 3 × CH_3), 1.91–1.96 (2H, m, CH_2), 3.66 (2H, t, $J = 5.2$ Hz, CH_2), 3.88 (2H, t, $J = 5.7$ Hz, CH_2), 7.97 (1H, dd, $J = 9.7$, 2.3 Hz, Ar), 8.01 (1H, d, $J = 2.3$ Hz, Ar), 8.39 (1H, d, $J = 9.2$ Hz, Ar); ^{13}C-NMR (125 MHz, $CDCl_3$) δ: 21.7, 30.0 (3C), 45.3, 45.5, 54.5, 119.9, 120.3, 130.0, 131.1, 132.8, 136.1, 146.5, 148.5; HRMS (FAB) *m/z* calcd for $C_{15}H_{19}N_4O_2S$ [M + H]$^+$ 319.1229; found: 319.1229.

Compound 25e. Using the general procedure as described for **25a**, compound **22e** (47.8 mg, 0.15 mmol) was allowed to react for 1 h with TFA (1.5 mL) and MS4Å (225 mg). Purification by flash chromatography over aluminum oxide with n-hexane–EtOAc (19:1 to 1:1) gave the title compound **25e** as pale yellow solid (24.9 mg, 63 %): mp 170–172 °C (from $CHCl_3$–n-hexane); IR (neat) cm^{-1}: 1620 (C=N), 1587 (NO_2), 1568 (C=N), 1523 (NO_2); ^{1}H-NMR (400 MHz, $CDCl_3$) δ: 1.97–2.03 (2H, m, CH_2), 3.74 (2H, t, $J = 5.6$ Hz, CH_2), 4.04 (2H, t, $J = 6.2$ Hz, CH_2), 7.41 (1H, br s, NH), 7.93 (1H, d, $J = 2.2$ Hz, Ar), 8.00 (1H, dd, $J = 9.0$, 2.2 Hz, Ar), 8.42 (1H, d, $J = 9.0$ Hz, Ar); ^{13}C-NMR (100 MHz, $CDCl_3$) δ: 20.8, 43.8, 45.2, 118.9, 120.5, 130.4, 130.8, 131.7, 145.1, 148.7, 151.3; *Anal.* calcd for $C_{11}H_{10}N_4O_2S$: C, 50.37; H, 3.84; N, 21.36. Found: C, 50.29; H, 4.03; N, 21.08.

3.1.12 Synthesis of 3,4-Dihydro-9-methoxy-2H,6H-pyrimido[1,2-c][1,3]benzothiazin-6-imine (25f)

3,4-Dihydro-9-methoxy-2H,6H-pyrimido[1,2-c][1,3]benzothiazine-6-thione   **21f** (66.1 mg, 0.25 mmol) was suspended into a 0.1 M NaOH in MeOH-H$_2$O (9:1) (5 mL), and the mixture was stirred for 12 h under reflux. After concentration, the residue was suspended in anhydrous EtOH (1 mL). BrCN (53.0 mg, 0.50 mmol) was added under an Ar atmosphere, and the mixture was stirred for 2 h under reflux. The reaction was quenched with 2 N NaOH, and the whole was extracted with CHCl$_3$, and dried over MgSO$_4$. After concentration, the residue was purified by flash chromatography over aluminum oxide with n-hexane–EtOAc (9:1) to give the title compound **25f** as colorless solid (37.6 mg, 61 %): mp 106 °C (from CHCl$_3$–n-hexane); IR (neat) cm^{-1}: 1620 (C=N), 1572 (C=N); ^{1}H-NMR (500 MHz, CDCl$_3$) δ: 1.94–1.98 (2H, m, CH$_2$), 3.66 (2H, t, J = 5.7 Hz, CH$_2$), 3.81 (3H, s, CH$_3$), 4.01 (2H, t, J = 6.0 Hz, CH$_2$), 6.50 (1H, d, J = 2.3 Hz, Ar), 6.76 (1H, dd, J = 9.0, 2.3 Hz, Ar), 7.15 (1H, br s, NH), 8.15 (1H, d, J = 9.0 Hz, Ar); ^{13}C-NMR (125 MHz, CDCl$_3$) δ: 21.0, 43.8, 44.8, 55.5, 107.3, 113.3, 119.5, 130.2, 130.6, 146.2, 153.4, 161.2; *Anal.* calcd for C$_{12}$H$_{13}$N$_3$OS: C, 58.28; H, 5.30; N, 16.99. Found: C, 58.15; H, 5.23; N, 16.79.

3.1.13 Synthesis of 3,4-Dihydro-9-methyl-2H, 6H-pyrimido[1,2-c][1,3]benzothiazin-6-imine (25g)

3,4-Dihydro-9-methyl-2H,6H-pyrimido[1,2-c][1,3]benzothiazine-6-thione   **21g** (62.1 mg, 0.25 mmol) was subjected to the general procedure as described for **25f** to give the title compound **25g** as colorless solid (39.2 mg, 68 %): mp 121 °C (from CHCl$_3$–n-hexane); IR (neat) cm^{-1}: 1620 (C=N), 1569 (C=N); ^{1}H-NMR (500 MHz, CDCl$_3$) δ: 1.94–1.99 (2H, m, CH$_2$), 2.32 (3H, s, CH$_3$), 3.67 (2H, t, J = 5.7 Hz, CH$_2$), 4.01 (2H, t, J = 6.3 Hz, CH$_2$), 6.84 (1H, s, Ar), 7.02 (1H, d, J = 8.6 Hz, Ar), 7.16 (1H, br s, NH), 8.10 (1H, d, J = 8.6 Hz, Ar); ^{13}C-NMR (125 MHz, CDCl$_3$) δ: 21.1, 21.1, 43.8, 44.9, 123.6, 124.1, 127.4, 128.6, 128.8, 141.1, 146.6, 153.6; HRMS (FAB): m/z calcd for C$_{12}$H$_{14}$N$_3$S [M + H]$^+$ 232.0908; found: 232.0912.

3.1.14 Synthesis of 9-(n-Butyl)-3,4-dihydro-2H, 6H-pyrimido[1,2-c][1,3]benzothiazin-6-imine (25h)

9-(*n*-Butyl)-*N*-(*tert*-butyl)-3,4-dihydro-2*H*,6*H*-pyrimido[1,2-*c*][1,3]benzothiazin-6-imine (22h). To a mixture of *N*-(*tert*-butyl)-9-bromo-3,4-dihydro-2H,6H-pyrimido[1,2-c][1,3]benzothiazin-6-imine **22k** (352.3 mg, 1.0 mmol), *n*-butylboronic acid (152.9 mg, 1.5 mmol), $Pd_2(dba)_3 \cdot CHCl_3$ (51.8 mg, 0.05 mmol) and Cs_2CO_3 (391.0 mg, 1.2 mmol) in 1,4-dioxane (2.5 mL) was added P(*tert*-Bu)$_3$ (0.024 mL, 0.1 mmol) under an Ar atmosphere, the mixture was stirred for 19 h under reflux. The mixture was filtered through a Celite pad, and concentrated. The residue was purified by flash chromatography over aluminum oxide with *n*-hexane–EtOAc (1:0 to 19:1) to give the title compound **22h** as a colorless oil (21.0 mg, 6 %): IR (neat) cm^{-1}: 1593 (C=N); ^{1}H-NMR (400 MHz, CDCl$_3$) δ; 0.91 (3H, t, $J = 7.3$ Hz, CH$_3$), 1.29-1.36 (2H, m, CH$_2$), 1.38 (9H, s, 3 × CH$_3$), 1.54-1.62 (2H, m, CH$_2$), 1.87–1.93 (2H, m, CH$_2$), 2.57 (2H, t, $J = 7.7$ Hz, CH$_2$), 3.60 (2H, t, $J = 5.4$ Hz, CH$_2$), 3.86 (2H, t, $J = 6.0$ Hz, CH$_2$), 6.91 (1H, d, $J = 1.2$ Hz, Ar), 7.01 (1H, dd, $J = 8.3$, 1.2 Hz, Ar), 8.08 (1H, d, $J = 8.3$ Hz, Ar); ^{13}C-NMR (100 MHz, CDCl$_3$) δ: 13.8, 21.9, 22.2, 30.0 (3C), 33.0, 35.2, 45.0, 45.4, 54.1, 123.9, 125.3, 126.5, 128.3, 128.7, 138.6, 145.4, 147.9; HRMS (FAB): *m/z* calcd for $C_{19}H_{28}N_3S$ [M + H]$^+$ 330.2004; found: 330.1999.

Compound 25h. Using the general procedure as described for **25a**, compound **22h** (10.3 mg, 0.03 mmol) was allowed to react for 1 h with TFA (1.0 mL) and MS4 Å (150 mg). Purification by flash chromatography over aluminum oxide with *n*-hexane–EtOAc (9:1) gave the title compound **25h** as colorless solid (5.2 mg, 61 %): mp 52–55 °C (from *n*-hexane); IR (neat) cm^{-1}: 1621 (C=N), 1571 (C=N); ^{1}H-NMR (400 MHz, CDCl$_3$) δ; 0.91 (3H, t, $J = 7.3$ Hz, CH$_3$), 1.29–1.38 (2H, m, CH$_2$), 1.54–1.61 (2H, m, CH$_2$), 1.93–1.99 (2H, m, CH$_2$), 2.58 (2H, t, $J = 7.7$ Hz, CH$_2$), 3.67 (2H, t, $J = 5.6$ Hz, CH$_2$), 4.01 (2H, t, $J = 6.2$ Hz, CH$_2$), 6.84 (1H, d, $J = 1.5$ Hz, Ar), 7.03–7.05 (1H, m, Ar), 8.11 (1H, d, $J = 8.3$ Hz, Ar); ^{13}C-NMR (100 MHz, CDCl$_3$) δ: 13.8, 21.1, 22.2, 33.0, 35.2, 43.8, 44.9, 123.0, 124.4, 126.8, 128.6, 128.8, 146.1, 146.6, 153.7; HRMS (FAB): *m/z* calcd for $C_{15}H_{20}N_3S$ [M + H]$^+$ 274.1378; found: 274.1372.

3.1.15 Synthesis of 9-Fluoro-3,4-dihydro-2H, 6H-pyrimido[1,2-c][1,3]benzothiazin-6-imine (25i)

9-Fluoro-3,4-dihydro-2*H*,6*H*-pyrimido[1,2-*c*][1,3]benzothiazine-6-thione   **21i** (63.1 mg, 0.25 mmol) was subjected to the general procedure as described for **25f** to give the title compound **25i** as colorless solid (30.4 mg, 52 %): mp 123–124 °C (from CHCl$_3$–*n*-hexane); IR (neat) cm^{-1}: 1624 (C=N), 1585 (C=N); ^{1}H-NMR (500 MHz, CDCl$_3$) δ: 1.94–2.00 (2H, m, CH$_2$), 3.67 (2H, t, $J = 5.7$ Hz, CH$_2$), 4.01 (2H, t, $J = 6.3$ Hz, CH$_2$), 6.75 (1H, dd, $J = 8.0$, 2.9 Hz, Ar), 6.91 (1H, ddd, $J = 8.6$, 8.0, 2.9 Hz, Ar), 7.22 (1H, br s, NH), 8.24 (1H, dd, $J = 8.6$, 5.7 Hz, Ar); ^{13}C-NMR (125 MHz, CDCl$_3$) δ: 21.0, 43.8, 44.8, 110.0 (d, $J = 25.2$ Hz), 113.9 (d, $J = 21.6$ Hz), 123.1, 130.9 (d, $J = 8.4$ Hz), 131.5 (d, $J = 8.4$ Hz), 146.4 (d, $J = 155.9$ Hz), 152.6, 163.7 (d, $J = 254.3$ Hz); ^{19}F-NMR (500 MHz, CDCl$_3$) δ: −109.1; *Anal.* calcd for $C_{11}H_{10}FN_3S$: C, 56.15; H, 4.28; N, 17.86. Found: C, 56.13; H, 4.44; N, 17.78.

3.1.16 Synthesis of 3,4-Dihydro-9-trifluoromethyl-2H, 6H-pyrimido[1,2-c][1,3]benzothiazin-6-imine (25j)

2-(2-Fluoro-4-trifluoromethylphenyl)-1,4,5,6-tetrahydropyrimidine (18j). To a solution of 2-fluoro-4-trifluoromethylbenzaldehyde **15j** (1.00 g, 5.21 mmol) in *t*-BuOH (49 mL) was added propylenediamine (424.7 mg, 5.73 mmol). The mixture was stirred at 70 °C for 30 min, and then K_2CO_3 (2.16 g, 15.6 mmol) and I_2 (1.65 g, 6.51 mmol) were added. After being stirred at same temperature for 3 h, the mixture was quenched with sat. Na_2SO_3. The organic layer was separated and concentrated. The resulting solid was dissolved with H_2O, and then pH was adjusted to 12–14 with 2 N NaOH. The whole was extracted with $CHCl_3$, and the extract was dried over $MgSO_4$. After concentration, the resulting solid was re-crystallized from $CHCl_3$–*n*-hexane to give the title compound **18j** as colorless crystals (0.84 g, 65 %): mp 108–110 °C; IR (neat) cm^{-1}: 1620 (C=N); ^{1}H-NMR (500 MHz, $CDCl_3$) δ: 1.86–1.90 (2H, m, CH_2), 3.52 (4H, t, $J = 5.2$ Hz, 2 × CH_2), 5.34 (1H, br s, NH), 7.33 (1H, d, $J = 11.5$ Hz, Ar), 7.42 (1H, d, $J = 8.6$ Hz, Ar), 7.96 (1H, dd, $J = 8.6, 8.0$ Hz, Ar); ^{13}C-NMR (100 MHz, $CDCl_3$) δ: 20.5, 42.2 (2C), 113.4 (dq, $J = 26.9, 3.9$ Hz), 120.9–121.0 (m), 123.0 (dq, $J = 273.0, 2.5$ Hz), 128.0 (d, $J = 13.2$ Hz), 131.5 (d, $J = 4.1$ Hz), 132.8 (dq, $J = 33.7, 9.1$ Hz), 150.4, 159.6 (d, $J = 249.1$ Hz); ^{19}F-NMR (500 MHz, $CDCl_3$) δ: -115.1, -63.4; *Anal.* calcd for $C_{11}H_{10}F_4N_2$: C, 53.66; H, 4.09; N, 11.38. Found: C, 53.82; H, 4.06; N, 11.43.

N-(*tert*-Butyl)-3,4-dihydro-9-trifluoromethyl-2H, 6H-pyrimido[1,2-c][1,3] benzothiazin-6-imine (22j). Using the general procedure as described for **22e**, compound **18j** (246.2 mg, 1.0 mmol) was allowed to react at 80 °C for 2 h. Purification by flash chromatography over aluminum oxide with *n*-hexane–EtOAc (1:0 to 9:1) gave the title compound **22j** as colorless solid (219.4 mg, 64 %): mp 82 °C (from *n*-hexane); IR (neat) cm^{-1}: 1601 (C=N), 1569 (C=N); ^{1}H-NMR (500 MHz, $CDCl_3$) δ: 1.39 (9H, s, 3 × CH_3), 1.90–1.95 (2H, m, CH_2), 3.64 (2H, t, $J = 5.4$ Hz, CH_2), 3.88 (2H, t, $J = 6.3$ Hz, CH_2), 7.38 (1H, s, Ar), 7.41 (1H, d, $J = 8.6$ Hz, Ar), 8.31 (1H, d, $J = 8.6$ Hz, Ar); ^{13}C-NMR (100 MHz, $CDCl_3$) δ: 21.8, 29.9 (3C), 45.2, 45.4, 54.3, 121.6 (q, $J = 4.0$ Hz), 122.4 (q, $J = 3.6$ Hz), 123.5 (q, $J = 272.7$ Hz), 129.2, 130.1, 130.7, 132.0 (q, $J = 33.2$ Hz), 136.9, 146.9; ^{19}F-NMR (500 MHz, $CDCl_3$) δ: -63.6; HRMS (FAB): *m/z* calcd for $C_{16}H_{19}F_3N_3S$ [M + H]$^+$ 342.1252; found: 342.1252.

Compound 25j. Using the general procedure as described for **25a**, compound **22j** (68.3 mg, 0.20 mmol) was allowed to react for 1 h. Purification by flash chromatography over aluminum oxide with *n*-hexane–EtOAc (8:2) gave the title compound **25j** as colorless solid (48.2 mg, 84 %): mp 91.5 °C (from $CHCl_3$–*n*-hexane); IR (neat) cm^{-1}: 1625 (C=N), 1561 (C=N); ^{1}H-NMR (500 MHz, $CDCl_3$) δ: 1.96–2.01 (2H, m, CH_2), 3.71 (2H, t, $J = 5.7$ Hz, CH_2), 4.03 (2H, t, $J = 6.3$ Hz, CH_2), 7.27 (2H, m, Ar, NH), 7.44 (1H, dd, $J = 8.3, 1.4$ Hz, Ar), 8.35 (1H, d, $J = 8.6$ Hz, Ar), ^{13}C-NMR (125 MHz, $CDCl_3$) δ: 20.9, 43.8, 45.0, 120.7 (q, $J = 4.0$ Hz), 122.7 (q, $J = 3.2$ Hz), 123.3 (q, $J = 272.7$ Hz), 129.6, 129.7,

129.9, 132.5 (q, $J = 33.2$ Hz), 145.6, 152.1; ^{19}F-NMR (500 MHz, CDCl$_3$) δ: -63.8; *Anal.* calcd for C$_{12}$H$_{10}$F$_3$N$_3$S: C, 50.52; H, 3.53; N, 14.73. Found: C, 50.51; H, 3.50; N, 14.69.

3.1.17 Synthesis of 9-Bromo-3,4-dihydro-2H,6H-pyrimido[1,2-c][1,3]benzothiazin-6-imine (25k)

2-(4-Bromo-2-fluorophenyl)-1,4,5,6-tetrahydropyrimidine (18k). 4-Bromo-2-fluorobenzaldehyde 15k (1.02 g, 5.0 mmol) was subjected to the general procedure as described for **18j** to give the title compound **18k** as colorless crystals (0.80 g, 62 %): mp 135–137 °C (from CHCl$_3$–*n*-hexane); IR (neat) cm^{-1}: 1622 (C=N); ^{1}H-NMR (400 MHz, CDCl$_3$) δ: 1.83–1.89 (2H, m, CH$_2$), 3.49 (4H, t, $J = 5.9$ Hz, 2 × CH$_2$), 4.88 (1H, br s, NH), 7.24 (1H, dd, $J = 11.2$, 2.0 Hz, Ar), 7.30 (1H, dd, $J = 8.5$, 2.0 Hz, Ar), 7.71 (1H, dd, $J = 8.5$, 8.3 Hz, Ar); ^{13}C-NMR (100 MHz, CDCl$_3$) δ: 20.6, 42.3 (2C), 119.5 (d, $J = 27.3$ Hz), 123.4 (d, $J = 3.3$ Hz), 123.6 (d, $J = 5.0$ Hz), 127.7 (d, $J = 3.3$ Hz), 131.6 (d, $J = 4.1$ Hz), 150.7, 159.8 (d, $J = 251.6$ Hz); ^{19}F-NMR (500 MHz, CDCl$_3$) δ: -114.7; *Anal.* calcd for C$_{10}$H$_{10}$BrFN$_2$: C, 46.72; H, 3.92; N, 10.90. Found: C, 46.66; H, 3.82; N, 10.87.

9-Bromo-*N*-(*tert*-butyl)-3,4-dihydro-2*H*,6*H*-pyrimido[1,2-*c*][1,3]benzothiazin-6-imine (22k). Using the general procedure as described for **22e**, compound **18k** (257.1 mg, 1.00 mmol) was allowed to react at rt overnight. Purification by flash chromatography over aluminum oxide with *n*-hexane–EtOAc (1:0 to 9:1) gave the title compound 22k as colorless solid (295.6 mg, 84 %): mp 107–108 °C (from *n*-hexane); IR (neat) cm^{-1}: 1596 (C=N); ^{1}H-NMR (400 MHz, CDCl$_3$) δ: 1.38 (9H, s, 3 × CH$_3$), 1.87–1.93 (2H, m, CH$_2$), 3.60 (2H, t, $J = 5.6$ Hz, CH$_2$), 3.85 (2H, t, $J = 6.1$ Hz, CH$_2$), 7.26–7.31 (2H, m, Ar), 8.05 (1H, d, $J = 8.5$ Hz, Ar); ^{13}C-NMR (100 MHz, CDCl$_3$) δ: 21.8, 30.0 (3C), 45.0, 45.4, 54.3, 124.4, 126.7, 126.8, 129.1, 130.1, 130.9, 137.2, 147.2; *Anal.* calcd for C$_{15}$H$_{18}$BrN$_3$S: C, 51.14; H, 5.15; N, 11.93. Found: C, 51.30; H, 5.07; N, 11.82.

Compound 25k. Using the general procedure as described for **25a**, compound **22k** (52.8 mg, 0.15 mmol) was allowed to react for 2 h with TFA (1.5 mL) and MS4Å (225 mg). Purification by flash chromatography over silica gel with *n*-hexane–EtOAc (2:1) gave the title compound **25k** as colorless solid (40.2 mg, 91 %): mp 104–105 °C (from CHCl$_3$–*n*-hexane); IR (neat) cm^{-1}: 1620 (C=N), 1569 (C=N); ^{1}H-NMR (400 MHz, CDCl$_3$) δ: 1.94–1.99 (2H, m, CH$_2$), 3.67 (2H, t, $J = 5.5$ Hz, CH$_2$), 4.00 (2H, t, $J = 6.0$ Hz, CH$_2$), 7.19–7.34 (3H, m, NH, Ar), 8.08 (1H, d, $J = 8.8$ Hz, Ar); ^{13}C-NMR (100 MHz, CDCl$_3$) δ: 20.9, 43.8, 44.9, 125.0, 125.6, 125.9, 129.5, 130.4, 130.7, 145.8, 152.4; *Anal.* calcd for C$_{11}$H$_{10}$BrN$_3$S: C, 44.61; H, 3.40; N, 14.19. Found: C, 44.37; H, 3.28; N, 13.93.

3.1.18 Synthesis of 3,4-Dihydro-9-phenyl-2H, 6H-pyrimido[1,2-c][1,3]benzothiazin-6-imine (25l)

N-(tert-Butyl)-3,4-dihydro-9-phenyl-2H,6H-pyrimido[1,2-c][1,3]benzothiazin-6-imine (22l). To a solution of N-(*tert*-butyl)-9-bromo-3,4-dihydro-2H,6H-pyrimido[1,2-c][1,3]benzothiazin-6-imine **22k** (52.8 mg, 0.15 mmol) and phenylboronic acid (21.9 mg, 0.18 mmol) in a mixture of toluene (1.5 mL), EtOH (0.9 mL) and 1 M aq. K_2CO_3 (1.5 mL) was added $Pd(PPh_3)_4$ (6.9 mg, 4 mol %) and $PdCl_2(dppf) \cdot CH_2Cl_2$ (3.7 mg, 3 mol %). After being stirred at reflux for 1 h, the mixture was extracted with $CHCl_3$. The extract was dried over $MgSO_4$ and concentrated. The residue was purified by flash chromatography over aluminum oxide with n-hexane–EtOAc (1:0 to 9:1) to give the title compound **22l** as colorless solid (44.8 mg, 85 %): mp 122.5–124 °C (from $CHCl_3$–n-hexane); IR (neat) cm^{-1}: 1592 (C=N); ^{1}H-NMR (500 MHz, $CDCl_3$) δ: 1.40 (9H, s, 3 × CH_3), 1.90–1.95 (2H, m, CH_2), 3.64 (2H, t, $J = 5.4$ Hz, CH_2), 3.89 (2H, t, $J = 6.0$ Hz, CH_2), 7.33–7.37 (2H, m, Ar), 7.41–7.44 (3H, m, Ar), 7.58 (2H, d, $J = 6.9$ Hz, Ar), 8.25 (1H, d, $J = 8.6$ Hz, Ar); ^{13}C-NMR (125 MHz, $CDCl_3$) δ: 21.9, 30.0 (3C), 45.1, 45.4, 54.2, 122.7, 124.8, 126.5, 127.0 (2C), 128.0, 128.8 (2C), 128.9, 129.5, 138.3, 139.4, 142.9, 147.7; HRMS (FAB): m/z calcd for $C_{21}H_{24}N_3S$ [M + H]$^+$ 350.1691; found: 350.1683.

Compound 25l. Using the general procedure as described for **25a**, compound **22l** (25.1 mg, 0.07 mmol) was allowed to react for 1 h with TFA (1.0 mL) and MS4Å (105 mg). Purification by flash chromatography over aluminum oxide with n-hexane–EtOAc (8:2) gave the title compound **25l** as pale yellow solid (19.4 mg, 92 %): mp 122–124 °C (from $CHCl_3$–n-hexane); IR (neat) cm^{-1}: 1619 (C=N), 1567 (C=N); ^{1}H-NMR (500 MHz, $CDCl_3$) δ: 1.97–2.02 (2H, m, CH_2), 3.72 (2H, t, $J = 5.4$ Hz, CH_2), 4.04 (2H, t, $J = 6.3$ Hz, CH_2), 7.25–7.26 (1H, m, Ar), 7.37–7.40 (1H, m, Ar), 7.43–7.47 (3H, m, Ar), 7.58 (2H, d, $J = 7.4$ Hz, Ar), 8.29 (1H, d, $J = 8.6$ Hz, Ar); ^{13}C-NMR (125 MHz, $CDCl_3$) δ: 21.1, 43.8, 45.0, 121.8, 121.8, 125.1, 125.5, 127.0 (2C), 128.2, 128.9 (2C), 129.4, 139.2, 143.5, 146.5, 153.4; HRMS (FAB): m/z calcd for $C_{17}H_{16}N_3S$ [M + H]$^+$ 294.1065; found: 294.1069.

3.1.19 Synthesis of 3,4-Dihydro-9-vinyl-2H, 6H-pyrimido[1,2-c][1,3]benzothiazin-6-imine (25m)

N-(tert-Butyl)-3,4-dihydro-9-vinyl-2H,6H-pyrimido[1,2-c][1,3]benzothiazin-6-imine (22m). Using the general procedure as described for 22l, N-(*tert*-butyl)-9-bromo-3,4-dihydro-2H,6H-pyrimido[1,2-c][1,3]benzothiazin-6-imine 22k (528.4 mg, 1.5 mmol) was allowed to react with vinylboronic acid pinacol ester (0.305 mL, 1.8 mmol) for 1 h. Purification by flash chromatography over aluminum oxide with n-hexane–EtOAc (1:0 to 9:1) gave the title compound 22m as colorless solid (455.7 mg, > 99 %): mp 67–68 °C (from *n*-hexane); IR (neat) cm^{-1}: 1589 (C=N); ^{1}H-NMR (400 MHz, CDCl$_3$) δ: 1.39 (9H, s, 3 × CH$_3$), 1.88–1.94 (2H, m, CH$_2$), 3.62 (2H, t, J = 5.6 Hz, CH$_2$), 3.87 (2H, t, J = 6.1 Hz, CH$_2$), 5.33 (1H, d, J = 11.0 Hz, CH), 5.79 (1H, d, J = 17.6 Hz, CH), 6.64 (1H, dd, J = 17.6, 11.0 Hz, CH), 7.12 (1H, d, J = 1.7 Hz, Ar), 7.23 (1H, dd, J = 8.3, 1.7 Hz, Ar), 8.14 (1H, d, J = 8.3 Hz, Ar); ^{13}C-NMR (100 MHz, CDCl$_3$) δ: 21.9, 30.0 (3C), 45.1, 45.4, 54.1, 115.9, 122.1, 123.7, 127.0, 128.6, 129.3, 135.4, 138.3, 139.3, 147.7; HRMS (FAB): *m/z* calcd for C$_{17}$H$_{22}$N$_3$S [M + H]$^+$ 300.1534; found: 300.1536.

Compound 25m. Using the general procedure as described for **25a**, compound **22m** (60.4 mg, 0.2 mmol) was allowed to react for 1 h. Purification by flash chromatography over aluminum oxide with *n*-hexane–EtOAc (8:2) gave the title compound **25m** as colorless solid (42.1 mg, 87 %): mp 76–77 °C (from CHCl$_3$–*n*-hexane); IR (neat) cm^{-1}: 1618 (C=N), 1564 (C=N); ^{1}H-NMR (400 MHz, CDCl$_3$) δ: 1.95-2.01 (2H, m, CH$_2$), 3.69 (2H, t, J = 5.4 Hz, CH$_2$), 4.02 (2H, t, J = 6.1 Hz, CH$_2$), 5.36 (1H, d, J = 10.9 Hz, CH), 5.81 (1H, d, J = 17.7 Hz, CH), 6.65 (1H, dd, J = 17.7, 10.9 Hz, CH), 7.04 (1H, s, Ar), 7.20 (1H, br s, NH), 7.26–7.28 (1H, m, Ar), 8.17 (1H, d, J = 8.5 Hz, Ar); ^{13}C-NMR (125 MHz, CDCl$_3$) δ: 21.0, 43.8, 44.9, 116.4, 121.1, 124.0, 125.8, 129.0 (2C), 135.2, 139.8, 146.4, 153.3; HRMS (FAB): *m/z* calcd for C$_{13}$H$_{14}$N$_3$S [M + H]$^+$ 244.0908; found: 244.0911.

3.1.20 Synthesis of 3,4-Dihydro-9-styryl-2H, 6H-pyrimido[1,2-c][1,3]benzothiazin-6-imine (25n)

N-(*tert*-Butyl)-3,4-dihydro-9-styryl-2H,6H-pyrimido[1,2-c][1,3]benzothiazin-6-imine (22n). Using the general procedure as described for **22l**, *N*-(*tert*-butyl)-9-bromo-3,4-dihydro-2*H*,6*H*-pyrimido[1,2-c][1,3]benzothiazin-6-imine **22k** (52.8 mg, 0.15 mmol) was allowed to react with styrylboronic acid pinacol ester (41.4 mg, 0.18 mmol) for 1 h. Purification by flash chromatography over aluminum oxide with *n*-hexane–EtOAc (1:0 to 9:1) gave the title compound **22n** as colorless solid (50.9 mg, 90 %): mp 124.5–125 °C (from CHCl$_3$–*n*-hexane); IR (neat) cm^{-1}: 1590 (C=N); ^{1}H-NMR (400 MHz, CDCl$_3$) δ: 1.40 (9H, s, 3 × CH$_3$), 1.88–1.94 (2H, m, CH$_2$), 3.63 (2H, t, J = 5.5 Hz, CH$_2$), 3.87 (2H, t, J = 6.1 Hz, CH$_2$), 7.01 (1H, d, J = 16.3 Hz, CH), 7.14 (1H, d, J = 16.3 Hz, CH), 7.22 (1H, d, J = 1.7 Hz, Ar), 7.27–7.38 (4H, m, Ar), 7.50 (2H, d, J = 7.3 Hz, Ar), 8.17 (1H, d, J = 8.3 Hz, Ar); ^{13}C-NMR (100 MHz, CDCl$_3$) δ: 21.9, 30.0 (3C), 45.1, 45.4, 54.2, 122.2, 124.0, 126.6, 126.7 (2C), 127.0, 128.1, 128.7 (2C), 128.8, 129.4, 130.7, 136.8, 138.3, 139.2, 147.7; HRMS (FAB): *m/z* calcd for C$_{23}$H$_{26}$N$_3$S [M + H]$^+$ 376.1847; found: 376.1845.

Compound 25n. Using the general procedure as described for **25a**, compound **22n** (31.7 mg, 0.084 mmol) was allowed to react for 2 h. Purification by flash chromatography over aluminum oxide with *n*-hexane–EtOAc (7:3) gave the title

compound **25n** as colorless solid (20.2 mg, 75 %): mp 111–113 °C (from CHCl$_3$–
n-hexane); IR (neat) cm^{-1}: 1618 (C=N), 1567 (C=N); ^{1}H-NMR (400 MHz,
CDCl$_3$) δ: 1.94–2.00 (2H, m, CH$_2$), 3.69 (2H, t, J = 5.6 Hz, CH$_2$), 4.02 (2H, t,
J = 6.2 Hz, CH$_2$), 7.00 (1H, d, J = 16.3 Hz, CH), 7.12–7.16 (2H, m, CH, Ar),
7.20 (1H, br s, NH), 7.26–7.30 (1H, m, Ar), 7.34–7.38 (3H, m, Ar), 7.50 (2H, d,
J = 7.6 Hz, Ar), 8.20 (1H, d, J = 8.5 Hz, Ar); ^{13}C-NMR (100 MHz, CDCl$_3$) δ:
21.0, 43.8, 45.0, 121.2, 124.2, 125.5, 126.6, 126.7 (2C), 128.2, 128.7 (2C), 129.1,
129.2, 131.1, 136.6, 139.7, 146.4, 153.3; *Anal.* calcd for C$_{19}$H$_{17}$N$_3$S: C, 71.44; H,
5.36; N, 13.15. Found: C, 71.17; H, 5.24; N, 13.07.

3.1.21   Synthesis of 3,4-Dihydro-9-pentenyl-2H, 6H-pyrimido[1,2-c][1,3]benzothiazin-6-imine (25o)

**N-(*tert*-Butyl)-3,4-dihydro-9-pentenyl-2H,6H-pyrimido[1,2-c][1,3]benzothia-
zin-6-imine (22o).** Using the general procedure as described for **22l**, N-(*tert*-butyl)-
9-bromo-3,4-dihydro-2H,6H-pyrimido[1,2-c][1,3]benzothiazin-6-imine   **22k**
(52.8 mg, 0.15 mmol) was allowed to react with pentenylboronic acid pinacol ester
(35.2 mg, 0.18 mmol) for 1 h. Purification by flash chromatography over aluminum oxide with *n*-hexane–EtOAc (1:0 to 9:1) gave the title compound **22o** as a
colorless oil (44.2 mg, 86 %): IR (neat) cm^{-1}: 1590 (C=N); ^{1}H-NMR (400 MHz,
CDCl$_3$) δ; 0.95 (t, J = 7.4 Hz, 3H, CH$_3$), 1.38 (9H, s, 3 × CH$_3$), 1.46–1.54 (2H, m,
CH$_2$), 1.87–1.93 (2H, m, CH$_2$), 2.16–2.21 (2H, m, CH$_2$), 3.61 (2H, t, J = 5.6 Hz,
CH$_2$), 3.86 (2H, t, J = 6.2 Hz, CH$_2$), 6.29–6.30 (2H, m, 2 × CH), 7.05 (1H, d,
J = 1.7 Hz, Ar), 7.17 (1H, dd, J = 8.3, 1.7 Hz, Ar), 8.10 (1H, d, J = 8.3 Hz, Ar);
^{13}C-NMR (100 MHz, CDCl$_3$) δ: 13.7, 21.9, 22.3, 30.0 (3C), 35.1, 45.1, 45.4, 54.1,
121.6, 123.6, 126.0, 128.5, 128.6, 129.1, 133.4, 138.5, 139.8, 147.8; HRMS (FAB):
m/z calcd for C$_{20}$H$_{28}$N$_3$S [M + H]$^+$ 342.2004; found: 342.2007.

Compound 25o. Using the general procedure as described for **25a**, compound
22o (40.0 mg, 0.12 mmol) was allowed to react for 2 h. Purification by flash
chromatography over aluminum oxide with *n*-hexane–EtOAc (8:2) gave the title
compound **25o** as a colorless oil (31.9 mg, 95 %): IR (neat) cm^{-1}: 1619 (C=N),
1568 (C=N); ^{1}H-NMR (400 MHz, CDCl$_3$) δ; 0.95 (3H, t, J = 7.4 Hz, CH$_3$),
1.45–1.54 (2H, m, CH$_2$), 1.93–1.99 (2H, m, CH$_2$), 2.17–2.22 (2H, m, CH$_2$), 3.68
(2H, t, J = 5.5 Hz, CH$_2$), 4.01 (2H, t, J = 6.2 Hz, CH$_2$), 6.29–6.31 (2H, m,
2 × CH), 6.96 (1H, d, J = 1.7 Hz, Ar), 7.16 (1H, br s, NH), 7.19 (1H, dd, J = 8.5,
1.7 Hz, Ar), 8.13 (1H, d, J = 8.5 Hz, Ar); ^{13}C-NMR (100 MHz, CDCl$_3$) δ: 13.7,
21.0, 22.3, 35.1, 43.8, 44.9, 120.6, 123.8, 124.9, 128.3, 128.9, 129.0, 133.9, 140.3,
146.5, 153.5; HRMS (FAB): *m/z* calcd for C$_{16}$H$_{20}$N$_3$S [M + H]$^+$ 286.1378;
found:286.1376.

3.1.22 Synthesis of 9-Azido-3,4-dihydro-2H, 6H-pyrimido[1,2-c][1,3]benzothiazin-6-imine (25p)

9-Azido-N-(*tert*-butyl)-3,4-dihydro-2H,6H-pyrimido[1,2-c][1,3]benzothiazin-6-imine (22p). To a solution of 9-amino-N-(*tert*-butyl)-3,4-dihydro-2H,6H-pyrimido[1,2-c][1,3]benzothiazin-6-imine **28** (100.9 mg, 0.35 mmol) in AcOH (2 mL) and H_2O (1 mL) was added $NaNO_2$ (33.8 mg, 0.49 mmol) at 0 °C, and the stirring was continued for 1 h. NaN_3 (34.1 mg, 0.53 mmol) was added to the reaction mixture, and stirring was continued for 30 min at rt. Reaction mixture was neutralized with K_2CO_3, and the whole was extracted with $CHCl_3$, and dried over $MgSO_4$. After concentration, the residue was purified by flash chromatography over aluminum oxide with n-hexane–EtOAc (9:1) to give the title compound **22p** as pale yellow solid (77.3 mg, 70 %): mp 79–80 °C (from n-hexane); IR (neat) cm^{-1}: 2104 (N_3), 1592 (C=N); ^{1}H-NMR (400 MHz, $CDCl_3$) δ: 1.38 (9H, s, 3 × CH_3), 1.88–1.94 (2H, m, CH_2), 3.60 (2H, t, $J = 5.6$ Hz, CH_2), 3.86 (2H, t, $J = 6.2$ Hz, CH_2), 6.74 (1H, d, $J = 2.3$ Hz, Ar), 6.84 (1H, dd, $J = 8.5, 2.3$ Hz, Ar), 8.19 (1H, d, $J = 8.5$ Hz, Ar); ^{13}C-NMR (100 MHz, $CDCl_3$) δ: 21.8, 30.0 (3C), 45.0, 45.4, 54.2, 114.2, 116.8, 124.5, 130.3, 130.9, 137.4, 142.0, 147.1; HRMS (FAB): *m/z* calcd for $C_{15}H_{19}N_6S$ [M + H]$^+$ 315.1392; found: 315.1398.

Compound 25p. Using the general procedure as described for **25a**, compound **22p** (77.3 mg, 0.25 mmol) was allowed to react for 2 h with TFA (3.5 mL) and MS4 Å (525 mg). Purification by recrystallization from MeOH–Et$_2$O gave the title compound **25p** as pale yellow crystals (27.0 mg, 42 %): mp 120–121 °C; IR (neat) cm^{-1}: 2107 (N_3), 1615 (C=N), 1569 (C=N); ^{1}H-NMR (400 MHz, DMSO-d_6) δ: 1.82–1.88 (2H, m, CH_2), 3.56 (2H, t, $J = 5.5$ Hz, CH_2), 3.89 (2H, t, $J = 5.4$ Hz, CH_2), 6.97 (1H, dd, $J = 8.8, 2.4$ Hz, Ar), 7.03 (1H, d, $J = 2.4$ Hz, Ar), 8.17 (1H, d, $J = 8.8$ Hz, Ar), 8.76 (1H, s, NH); ^{13}C-NMR (100 MHz, DMSO-d_6) δ: 20.6, 43.1, 44.2, 113.7, 117.0, 122.6, 130.2, 130.8, 141.9, 144.7, 150.0; *Anal.* calcd for $C_{11}H_{10}N_6S$: C, 51.15; H, 3.90; N, 32.54. Found: C, 51.07; H, 3.88; N, 32.28.

3.1.23 Synthesis of 9-(4-Benzoylphenyl)-3,4-dihydro-2H, 6H-pyrimido[1,2-c][1,3]benzothiazin-6-imine (25q)

9-(4-Benzoylphenyl)-N-(*tert*-butyl)-3,4-dihydro-2H, 6H-pyrimido[1,2-c][1,3] benzothiazin-6-imine (22q). Using the general procedure as described for **22l**, N-(*tert*-butyl)-9-bromo-3,4-dihydro-2H,6H-pyrimido[1,2-c][1,3]benzothiazin-6-imine **22k** (52.8 mg, 0.15 mmol) was allowed to react with 4-benzoylphenylboronic acid (40.7 mg, 0.18 mmol) for 1 h. Purification by flash chromatography over aluminum oxide with n-hexane–EtOAc (8:2) gave the title compound **22q** as colorless solid (55.6 mg, 82 %): mp 187–189 °C (from $CHCl_3$–n-hexane); IR (neat) cm^{-1}: 1656 (C=O), 1593 (C=N); ^{1}H-NMR (400 MHz, $CDCl_3$) δ: 1.41 (9H, s, 3 × CH_3), 1.91–1.97 (2H, m, CH_2), 3.65 (2H, t, $J = 5.5$ Hz, CH_2), 3.90 (2H, t, $J = 6.1$ Hz, CH_2), 7.39 (1H, d, $J = 1.7$ Hz, Ar), 7.46–7.53 (3H, m, Ar), 7.60 (1H, t, $J = 7.4$ Hz, Ar), 7.70 (2H, d, $J = 8.0$ Hz, Ar), 7.82 (2H, d, $J = 7.3$ Hz, Ar), 7.88 (2H, d, $J = 8.0$ Hz, Ar), 8.30 (1H, d, $J = 8.3$ Hz, Ar); ^{13}C-NMR (100 MHz, CDCl$_3$) δ: 21.9, 30.0 (3C), 45.2, 45.4, 54.2, 123.0, 124.8, 126.9 (2C), 127.4, 128.3 (2C), 129.1, 129.9, 130.0 (2C), 130.7 (2C), 132.5, 136.9, 137.6, 137.9, 141.7, 143.3, 147.6, 196.1; HRMS (FAB): m/z calcd for C$_{28}$H$_{28}$N$_3$OS [M + H]$^+$ 454.1953; found: 454.1954.

Compound 25q. Using the general procedure as described for **25a**, compound **22q** (30.4 mg, 0.067 mmol) was allowed to react for 1 h with TFA (1.0 mL) and MS4 Å (150 mg). Purification by flash chromatography over aluminum oxide with *n*-hexane–EtOAc (7:3) gave the title compound **25q** as colorless solid (16.7 mg, 63 %): mp 155–156 °C (from CHCl$_3$–*n*-hexane); IR (neat) cm^{-1}: 1655 (C=O), 1619 (C=N), 1561 (C=N); ^{1}H-NMR (400 MHz, CDCl$_3$) δ: 1.97–2.03 (2H, m, CH$_2$), 3.72 (2H, t, $J = 5.5$ Hz, CH$_2$), 4.05 (2H, t, $J = 6.1$ Hz, CH$_2$), 7.30 (1H, d, $J = 1.7$ Hz, Ar), 7.48–7.52 (3H, m, Ar), 7.59–7.63 (1H, m, Ar), 7.68 (2H, d, $J = 8.3$ Hz, Ar), 7.81–7.83 (2H, m, Ar), 7.89 (2H, d, $J = 8.3$ Hz, Ar), 8.32 (1H, d, $J = 8.5$ Hz, Ar); ^{13}C-NMR (100 MHz, CDCl$_3$) δ: 21.0, 43.8, 45.0, 122.0, 125.1, 126.2, 126.9 (2C), 128.3 (2C), 129.5, 129.6, 130.0 (2C), 130.7 (2C), 132.5, 137.1, 137.5, 142.2, 143.0, 146.3, 153.0, 196.0; HRMS (FAB): m/z calcd for C$_{20}$H$_{18}$N$_3$OS [M + H]$^+$ 398.1327; found: 398.1333.

3.1.24 Synthesis of 10-(N,N-Dimethylamino)-3,4-dihydro-2H,6H-pyrimido[1,2-c][1,3]benzothiazin-6-imine (26a)

N-(*tert*-Butyl)-10-(N,N-dimethylamino)-3,4-dihydro-2H,6H-pyrimido[1,2-c][1, 3]benzothiazin-6-imine (23a). To a mixture of 10-bromo-*N*-(*tert*-butyl)-3,4-dihydro-2*H*,6*H*-pyrimido[1,2-*c*][1,3]benzothiazin-6-imine **23k** (600.2 mg, 1.70 mmol) and Pd(P*t*-Bu$_3$)$_2$ (174.2 mg, 0.341 mmol) and KO*t*-Bu (573.3 mg, 5.11 mmol) in toluene (1.7 mL) was added 2.0 M Me$_2$NH in THF (2.55 mL, 5.11 mmol). The reaction was heated using a microwave reactor (standard mode) for 10 min at 170 °C. The whole was extracted with EtOAc. The extract was washed with brine, and dried over Na$_2$SO$_4$. After concentration, the residue was purified by flash chromatography over silica gel with *n*-hexane–EtOAc (6:4 to 5:5) to give the title compound **23a** as pale yellow solid (363.3 mg, 67.4 %): mp 86.1 °C; IR (neat) cm^{-1}: 1583 (C=N); ^{1}H-NMR (300 MHz, CDCl$_3$) δ: 1.38 (9H, s, 3 × CH$_3$), 1.86–1.94 (2H, m, CH$_2$), 2.97 (6H, s, 2 × CH$_3$), 3.61 (2H, t, $J = 5.3$ Hz, CH$_2$), 3.86 (2H, t, $J = 6.3$ Hz, CH$_2$), 6.78 (1H, dd, $J = 9.0$, 3.0 Hz, Ar), 6.98 (1H, d, $J = 8.4$ Hz, Ar), 7.56 (1H, d, $J = 2.4$ Hz, Ar); ^{13}C-NMR (75 MHz, CDCl$_3$) δ: 22.0, 29.9 (3C), 40.8 (2C), 45.1, 45.5, 54.0, 111.6, 115.4, 115.8, 125.3, 128.7, 139.7, 148.8, 149.3; HRMS (FAB): m/z calcd for C$_{17}$H$_{25}$N$_4$S [M + H]$^+$ 317.1800; found: 317.1796

Compound 26a. TFA (0.63 mL) was added to a mixture of **23a** (20 mg, 0.063 mmol) and MS4Å (110 mg, powder, activated by heating with Bunsen burner) in small amount of CHCl$_3$. After being stirred under reflux for 40 min, the mixture was added dropwise to Et$_3$N at 0 °C to adjust pH to 8–9. The whole was extracted with EtOAc. The extract was washed with sat. NaHCO$_3$, brine, and dried over Na$_2$SO$_4$. After concentration, the residue was purified by preparative TLC over aluminum oxide with *n*-hexane–EtOAc (7:3) to give the title compound **26a** as yellow solid (11.4 mg, 68.3 %): mp 134.5 °C; IR (neat) cm^{-1}: 1617 (C=N), 1552 (C=N); ^{1}H-NMR (300 MHz, CDCl$_3$) δ: 1.92–2.00 (2H, m, CH$_2$), 2.97 (6H, s, 2 × CH$_3$), 3.68 (2H, t, J = 5.7 Hz, CH$_2$), 4.01 (2H, t, J = 6.3 Hz, CH$_2$), 6.79 (1H, dd, J = 8.7, 3.3 Hz, Ar), 6.89 (1H, d, J = 8.7 Hz, Ar), 7.08 (1H, br s, NH), 7.58 (1H, d, J = 2.7 Hz, Ar); ^{13}C-NMR (75 MHz, CDCl$_3$) δ: 20.8, 40.6 (2C), 44.1, 44.3, 111.8, 114.6, 116.3, 124.6, 126.3, 148.7, 149.4, 154.4; *Anal.* calcd for C$_{13}$H$_{16}$N$_4$S: C, 59.97; H, 6.19; N, 21.52. Found: C, 59.91; H, 6.19; N, 21.41.

3.1.25 Synthesis of 3,4-Dihydro-10-nitro-2H, 6H-pyrimido[1,2-c][1,3]benzothiazin-6-imine (26e)

N-(tert-Butyl)-3,4-dihydro-10-nitro-2H,6H-pyrimido[1,2-c][1,3]benzothiazin-6-imine (23e). To a mixture of 2-(2-bromo-5-nitrophenyl)-1,4,5,6-tetrahydropyrimidine **19e** (50 mg, 0.209 mmol) in DMAc (0.70 mL) were added *tert*-butylisothiocyanate (0.053 mL, 0.418 mmol) and KO*t*-Bu (46.9 mg, 0.418 mmol) at 0 °C under an N$_2$ atmosphere. After being stirred at 0 °C for 1 h, sat. NH$_4$Cl was added. The whole was extracted with EtOAc. The extract was washed with brine, and dried over Na$_2$SO$_4$. After concentration, the residue was purified by flash chromatography over silica gel with *n*-hexane–EtOAc (1:0 to 9:1) to give the title compound **23e** as pale yellow solid (39.1 mg, 58.9 %): mp 123.8 °C; IR (neat) cm^{-1}: 1593 (NO$_2$), 1520 (NO$_2$); ^{1}H-NMR (300 MHz, CDCl$_3$) δ: 1.40 (s, 9H, 3 × CH$_3$), 1.90–1.98 (2H, m, CH$_2$), 3.67 (2H, t, J = 5.6 Hz, CH$_2$), 3.89 (2H, t, J = 6.3 Hz, CH$_2$), 7.23 (1H, m, Ar), 8.13 (1H, dd, J = 8.7, 2.7 Hz, Ar), 9.11 (1H, d, J = 2.7 Hz, Ar); ^{13}C-NMR (75 MHz, CDCl$_3$) δ: 21.7, 30.0 (3C), 45.1, 45.5, 54.5, 124.1, 124.3, 125.3, 128.5, 135.6, 137.1, 145.8, 146.1; HRMS (FAB): *m/z* calcd for C$_{15}$H$_{19}$N$_4$O$_2$S [M + H]$^+$ 319.1229; found: 319.1232.

Compound 26e. TFA (3.2 mL) was added to compound **23e** (100 mg, 0.314 mmol). After being stirred under reflux for 1.5 h, the mixture was added dropwise to Et$_3$N at 0 °C to adjust pH to 8–9. The whole was extracted with EtOAc. The extract was washed with sat. NaHCO$_3$, brine, and dried over Na$_2$SO$_4$. After concentration, the residue was purified by preparative TLC over aluminum oxide with *n*-hexane–EtOAc (7:3) to give the title compound 26e as orange solid (15.9 mg, 19.3 %): mp 167.9 °C; IR (neat) cm^{-1}: 1614 (C=N), 1576 (NO$_2$), 1557 (C=N), 1519 (NO$_2$); ^{1}H-NMR (300 MHz, CDCl$_3$) δ: 1.97–2.04 (2H, m, CH$_2$), 3.74 (2H, t, J = 5.6 Hz, CH$_2$), 4.05 (2H, t, J = 6.2 Hz, CH$_2$), 7.19 (1H, d, J = 9.0 Hz,

Ar), 7.38 (1H, br s, NH), 8.17 (1H, dd, $J = 8.7$, 2.7 Hz, Ar), 9.13 (1H, d, $J = 2.7$ Hz, Ar); ^{13}C-NMR (75 MHz, CDCl$_3$) δ: 20.8, 44.1, 44.9, 124.5 (2C), 124.8, 127.6, 136.7, 144.5, 146.3, 151.0; HRMS (FAB): m/z calcd for C$_{11}$H$_{11}$N$_4$O$_2$S [M + H]$^+$ 263.0603; found: 263.0606. The purity of the compound was 75 % by HPLC.

3.1.26 Synthesis of 3,4-Dihydro-10-methoxy-2H, 6H-pyrimido[1,2-c][1,3]benzothiazin-6-imine (26f)

N-(*tert*-Butyl)-3,4-dihydro-10-methoxy-2*H*,6*H*-pyrimido[1,2-*c*][1,3]benzothiazin-6-imine (23f). To a mixture of N-(*tert*-butyl)-9-bromo-3,4-dihydro-2*H*,6*H*-pyrimido[1,2-*c*][1,3]benzothiazin-6-imine **22k** (500.3 mg, 1.42 mmol) and Na-OMe (767 mg, 14.2 mmol, 28 % solution in MeOH,) in DMF (2.5 mL) was added CuBr (20.4 mg, 0.142 mmol). The mixture was stirred at 110 °C for 2.5 h. The whole was extracted with CH$_2$Cl$_2$. The extract was washed with brine, and dried over Na$_2$SO$_4$. After concentration, the residue was purified by flash chromatography over silica gel with *n*-hexane–EtOAc (6:4 to 4:6) to give the titlecompound **23f** as colorless solid (171.5 mg, 39.8 %): mp 87.1 °C; IR (neat) cm^{-1}: 1588 (C=N); ^{1}H-NMR (300 MHz, CDCl$_3$) δ: 1.38 (9H, s, 3 × CH$_3$), 1.87–1.95 (2H, m, CH$_2$), 3.62 (2H, t, $J = 5.6$ Hz, CH$_2$), 3.86 (5H, m, CH$_3$, CH$_2$), 6.92 (1H, dd, $J = 8.7$, 2.7 Hz, Ar), 7.00 (1H, d, $J = 8.4$ Hz, Ar), 7.75 (1H, d, $J = 3.3$ Hz, Ar); ^{13}C-NMR (75 MHz, CDCl$_3$) δ: 21.9, 29.9 (3C), 45.1, 45.5, 54.1, 55.6, 111.3, 118.9, 120.2, 125.7, 128.9, 138.8, 148.1, 158.1; *Anal.* calcd for C$_{16}$H$_{21}$N$_3$OS: C, 63.33; H, 6.98; N, 13.85. Found: C, 63.04; H, 6.97; N, 13.68.

Compound 26f. TFA (0.88 mL) was added to compound **23f** (26.7 mg, 0.088 mmol). After being stirred under reflux for 3 h, the mixture was added dropwise to Et$_3$N at 0 °C to adjust pH to 8–9. The whole was extracted with EtOAc. The extract was washed with sat. NaHCO$_3$, brine, and dried over Na$_2$SO$_4$. After concentration, the residue was purified by preparative TLC over aluminum oxide with *n*-hexane–EtOAc (7:3) to give the title compound **26f** as colorless solid (9.6 mg, 44 %): mp 89.0 °C; IR (neat) cm^{-1}: 1614 (C=N), 1562 (C=N); ^{1}H-NMR (300 MHz, CDCl$_3$) δ: 1.93–2.00 (2H, m, CH$_2$), 3.69 (2H, t, $J = 5.4$ Hz, CH$_2$), 3.85 (3H, s, CH$_3$), 4.02 (2H, t, $J = 6.2$ Hz, CH$_2$), 6.92–6.98 (2H, m, Ar), 7.15 (1H, br s, NH), 7.78 (1H, s, Ar); ^{13}C-NMR (75 MHz, CDCl$_3$) δ: 21.0, 43.9, 44.9, 55.6, 111.9, 119.3, 119.9, 124.8, 127.9, 146.7, 153.9, 158.3; *Anal.* calcd for C$_{12}$H$_{13}$N$_3$OS: C, 58.28; H, 5.30; N, 16.99. Found: C, 58.24; H, 5.36; N, 16.46. The purity of the compound was 92 % by HPLC.

3.1.27 Synthesis of 3,4-Dihydro-10-methyl-2H, 6H-pyrimido[1,2-c][1,3]benzothiazin-6-imine (26g)

2-(2-Fluoro-5-methylphenyl)-1,4,5,6-tetrahydropyrimidine (19g). 2-Fluoro-5-methylbenzaldehyde 16g (3.0 g, 21.7 mmol) was subjected to the general procedure as described for **18j** to give the title **19g** as colorless crystals (3.1 g, 75 %): mp 119–121 °C (from CHCl$_3$–n-hexane); IR (neat) cm^{-1}: 1626 (C=N); ^{1}H-NMR (500 MHz, CDCl$_3$) δ: 1.84–1.89 (2H, m, CH$_2$), 2.31 (3H, s, CH$_3$), 3.51 (4H, t, $J = 5.7$ Hz, 2 × CH$_2$), 5.01 (1H, s, NH), 6.92 (1H, dd, $J = 11.7$, 8.3 Hz, Ar), 7.09–7.12 (1H, m, Ar), 7.63 (1H, dd, $J = 7.4$, 2.3 Hz, Ar); ^{13}C-NMR (125 MHz, CDCl$_3$) δ: 20.5, 20.7, 42.3 (2C), 115.6 (d, $J = 24.0$ Hz), 123.7 (d, $J = 12.0$ Hz), 130.6 (d, $J = 3.6$ Hz), 131.3 (d, $J = 9.6$ Hz), 133.9 (d, $J = 3.6$ Hz), 151.7, 158.4 (d, $J = 244.7$ Hz); ^{19}F-NMR (500 MHz, CDCl$_3$) δ: −122.4; HRMS (FAB): m/z calcd for C$_{11}$H$_{14}$FN$_2$ [M + H]$^+$ 193.1141; found: 193.1140.

N-(tert-Butyl)-3,4-dihydro-10-methyl-2H,6H-pyrimido[1,2-c][1,3]benzothiazin-6-imine (23g). To a mixture of compound **19g** (0.50 g, 2.6 mmol) and KOt-Bu (0.58 g, 5.2 mmol) in DMAc (8.7 mL) was added tert-butylisothiocyanate (0.66 mL, 5.2 mmol) under an Ar atmosphere. After being stirred at 80 °C for 3 h, the whole was extracted with EtOAc. The whole was washed with sat. NaHCO$_3$, brine, and dried over MgSO$_4$. After concentration, the residue was purified by flash chromatography over silica gel with n-hexane–EtOAc (1:1) to give the title compound **23g** as colorless solid (0.21 g, 28 %): mp 76–77 °C (from n-hexane); IR (neat) cm^{-1}: 1597 (C=N); ^{1}H-NMR (400 MHz, CDCl$_3$) δ: 1.38 (9H, s, 3 × CH$_3$), 1.88–1.94 (2H, m, CH$_2$), 2.33 (3H, s, CH$_3$), 3.62 (2H, t, $J = 5.6$ Hz, CH$_2$), 3.87 (2H, t, $J = 6.2$ Hz, CH$_2$), 7.00 (1H, d, $J = 8.0$ Hz, Ar), 7.13 (1H, dd, $J = 8.0$, 1.3 Hz, Ar), 8.01 (1H, d, $J = 1.3$ Hz, Ar); ^{13}C-NMR (100 MHz, CDCl$_3$) δ: 21.0, 22.0, 29.9 (3C), 45.1, 45.4, 54.1, 124.4, 125.7, 127.6, 128.6, 131.1, 135.9, 138.7, 148.2; HRMS (FAB): m/z calcd for C$_{16}$H$_{22}$N$_3$S [M + H]$^+$ 288.1534; found: 288.1535.

Compound 26g. Using the general procedure as described for **25a**, compound **23g** (200 mg, 0.7 mmol) was allowed to react for 1 h with TFA (3.0 mL) and MS4Å (450 mg). Purification by preparative TLC over aluminum oxide with n-hexane–EtOAc (9:1) gave the title compound **26g** as colorless solid (150 mg, 92 %): mp 116 °C (from CHCl$_3$–n-hexane); IR (neat) cm^{-1}: 1623 (C=N), 1556 (C=N); ^{1}H-NMR (400 MHz, CDCl$_3$) δ: 1.94–2.00 (2H, m, CH$_2$), 2.34 (3H, s, CH$_3$), 3.69 (2H, t, $J = 5.6$ Hz, CH$_2$), 4.01 (2H, t, $J = 6.1$ Hz, CH$_2$), 6.94 (1H, d, $J = 8.0$ Hz, Ar), 7.15–7.17 (2H, m, Ar, NH), 8.04 (1H, s, Ar); ^{13}C-NMR (100 MHz, CDCl$_3$) δ: 21.0, 21.1, 43.8, 44.9, 123.5, 125.4, 126.5, 129.0, 131.6, 136.3, 146.9, 153.7; HRMS (FAB): m/z calcd for C$_{12}$H$_{14}$N$_3$S [M + H]$^+$ 232.0908; found: 232.0913.

3.1.28 Synthesis of 10-Bromo-3,4-dihydro-2H, 6H-pyrimido[1,2-c][1,3]benzothiazin-6-imine (26k)

2-(5-Bromo-2-fluorophenyl)-1,4,5,6-tetrahydropyrimidine (19k). 5-Bromo-2-fluorobenzaldehyde 16k (1.02 g, 5.0 mmol) was subjected to the general procedure as described for **18j** to give the title compound **19k** as colorless crystals (1.02 g, 79 %): mp 121–122 °C (from $CHCl_3$–n-hexane); IR (neat) cm^{-1}: 1623 (C=N); ^{1}H-NMR (400 MHz, $CDCl_3$) δ: 1.83–1.89 (2H, m, CH_2), 3.50 (4H, t, $J = 5.7$ Hz, 2 × CH_2), 5.28 (1H, br s, NH), 6.94 (1H, dd, $J = 11.1$, 8.8 Hz, Ar), 7.42 (1H, ddd, $J = 8.8$, 4.4, 2.7 Hz, Ar), 7.97 (1H, dd, $J = 6.8$, 2.7 Hz, Ar); ^{13}C-NMR (100 MHz, $CDCl_3$) δ: 20.6, 42.1 (2C), 117.1 (d, $J = 3.3$ Hz), 117.7 (d, $J = 25.7$ Hz), 126.2 (d, $J = 13.2$ Hz), 133.3 (d, $J = 3.3$ Hz), 133.6 (d, $J = 9.1$ Hz), 150.3, 159.1 (d, $J = 247.5$ Hz); ^{19}F-NMR (500 MHz, $CDCl_3$) δ: -119.5; *Anal.* calcd for $C_{10}H_{10}BrFN_2$: C, 46.72; H, 3.92; N, 10.90. Found: C, 46.59; H, 3.87; N, 10.89.

10-Bromo-N-(tert-butyl)-3,4-dihydro-2H,6H-pyrimido[1,2-c][1,3]benzothiazin-6-imine (23k). Using the general procedure as described for 22e, compound 19k (257.1 mg, 1.00 mmol) was allowed to react at rt overnight. Purification by flash chromatography over aluminum oxide with n-hexane–EtOAc (1:0 to 9:1) gave the title compound **23k** as colorless solid (111.6 mg, 32 %): mp 93–94 °C (from n-hexane); IR (neat) cm^{-1}: 1599 (C=N); ^{1}H-NMR (400 MHz, $CDCl_3$) δ: 1.38 (9H, s, 3 × CH_3), 1.88-1.93 (2H, m, CH_2), 3.62 (2H, t, $J = 5.6$ Hz, CH_2), 3.86 (2H, t, $J = 6.1$ Hz, CH_2), 6.97 (1H, d, $J = 8.5$ Hz, Ar), 7.41 (1H, dd, $J = 8.5$, 2.2 Hz, Ar), 8.36 (1H, d, $J = 2.2$ Hz, Ar); ^{13}C-NMR (100 MHz, $CDCl_3$) δ: 21.8, 29.9 (3C), 45.1, 45.4, 54.2, 119.7, 125.9, 128.1, 129.3, 131.2, 133.0, 137.4, 146.7; *Anal.* calcd for $C_{15}H_{18}BrN_3S$: C, 51.14; H, 5.15; N, 11.93. Found: C, 51.09; H, 4.98; N, 11.89.

Compound 26k. Using the general procedure as described for 25a, compound 23k (52.8 mg, 0.15 mmol) was allowed to react for 2 h with TFA (1.5 mL) and MS4Å (225 mg). Purification by flash chromatography over silica gel with n-hexane–EtOAc (2:1) gave the title compound **26k** as colorless crystals (39.7 mg, 89 %): mp 106–107 °C (from $CHCl_3$–n-hexane); IR (neat) cm^{-1}: 1621 (C=N), 1571 (C=N); ^{1}H-NMR (500 MHz, $CDCl_3$) δ: 1.94-1.99 (2H, m, CH_2), 3.69 (2H, t, $J = 5.4$ Hz, CH_2), 4.01 (2H, t, $J = 6.3$ Hz, CH_2), 6.91 (1H, d, $J = 8.6$ Hz, Ar), 7.20 (1H, br s, NH), 7.44 (1H, dd, $J = 8.6$, 2.3 Hz, Ar), 8.39 (1H, d, $J = 2.3$ Hz, Ar); ^{13}C-NMR (125 MHz, $CDCl_3$) δ: 20.9, 43.8, 44.9, 120.0, 125.0, 127.8, 128.3, 131.6, 133.5, 145.4, 152.5; *Anal.* calcd for $C_{11}H_{10}BrN_3S$: C, 44.61; H, 3.40; N, 14.19. Found: C, 44.51; H, 3.66; N, 14.06.

3.1.29 Synthesis of 3,4-Dihydro-10-phenyl-2H, 6H-pyrimido[1,2-c][1,3]benzothiazin-6-imine (26l)

N-(tert-Butyl)-3,4-dihydro-10-phenyl-2H,6H-pyrimido[1,2-c][1,3]benzothiazin-6-imine (23l). Using the general procedure as described for **22l**, 10-bromo-N-(tert-butyl)-3,4-dihydro-2H,6H-pyrimido[1,2-c][1,3]benzothiazin-6-imine **23k** (52.8 mg, 0.15 mmol) was allowed to react for 1 h. Purification by flash chromatography over aluminum oxide with n-hexane–EtOAc (1:0 to 9:1) gave the title compound **23l** as colorless solid (32.6 mg, 62 %): mp 101–103 °C (from n-hexane); IR (neat) cm^{-1}: 1594 (C=N); ^{1}H-NMR (500 MHz, CDCl$_3$) δ: 1.40 (9H, s, 3 × CH$_3$), 1.90–1.95 (2H, m, CH$_2$), 3.64 (2H, t, J = 5.4 Hz, CH$_2$), 3.89 (2H, t, J = 6.0 Hz, CH$_2$), 7.18 (1H, d, J = 8.0 Hz, Ar), 7.32 (1H, t, J = 7.4 Hz, Ar), 7.41 (2H, t, J = 7.4 Hz, Ar), 7.55 (1H, dd, J = 8.0, 2.0 Hz, Ar), 7.61 (2H, d, J = 7.4 Hz, Ar), 8.47 (1H, d, J = 2.0 Hz, Ar); ^{13}C-NMR (125 MHz, CDCl$_3$) δ: 21.9, 30.0 (3C), 45.1, 45.5, 54.2, 125.0, 126.9, 127.0 (2C), 127.4, 127.9, 128.1, 128.7 (2C), 128.7, 138.2, 139.1, 140.0, 147.9; HRMS (FAB): m/z calcd for C$_{21}$H$_{24}$N$_3$S [M + H]$^+$ 350.1691; found:350.1683.

Compound 26l. Using the general procedure as described for **25a**, compound **23l** (13.1 mg, 0.037 mmol) was allowed to react for 2 h with TFA (1.0 mL) and MS4Å (150 mg). Purification by flash chromatography over aluminum oxide with n-hexane–EtOAc (9:1) gave the title compound **26l** as colorless solid (8.4 mg, 77 %): mp 82 °C (from CHCl$_3$–n-hexane); IR (neat) cm^{-1}: 1621 (C=N), 1550 (C=N); ^{1}H-NMR (400 MHz, CDCl$_3$) δ: 1.95-2.01 (2H, m, CH$_2$), 3.71 (2H, t, J = 5.6 Hz, CH$_2$), 4.03 (2H, t, J = 6.2 Hz, CH$_2$), 7.10–7.44 (5H, m, Ar), 7.56–7.64 (3H, m, Ar), 8.50 (1H, d, J = 2.2 Hz, Ar); ^{13}C-NMR (100 MHz, CDCl$_3$) δ: 21.1, 43.9, 45.0, 124.0, 126.9, 127.1 (2C), 127.3, 127.6, 128.8 (2C), 129.2, 139.4, 139.8, 146.6, 152.1, 153.3; Anal. calcd for C$_{17}$H$_{15}$N$_3$S: C, 69.59; H, 5.15; N, 14.32. Found: C, 69.61; H, 5.13; N, 14.22.

3.1.30 Synthesis of 3,4-Dihydro-10-vinyl-2H, 6H-pyrimido[1,2-c][1,3]benzothiazin-6-imine (26m)

N-(tert-Butyl)-3,4-dihydro-10-vinyl-2H,6H-pyrimido[1,2-c][1,3]benzothiazin-6-imine (23m). Using the general procedure as described for **22l**, 10-bromo-N-(tert-butyl)-3,4-dihydro-2H,6H-pyrimido[1,2-c][1,3]benzothiazin-6-imine **23k** (52.8 mg, 0.15 mmol) was allowed to react with vinylboronic acid pinacol ester (0.031 mL, 0.18 mmol) for 1 h. Purification by flash chromatography over aluminum oxide with n-hexane–EtOAc (1:0 to 9:1) gave the title compound **23m** as a colorless oil (30.5 mg, 68 %): IR (neat) cm^{-1}: 1595 (C=N); ^{1}H-NMR (400 MHz, CDCl$_3$) δ: 1.38 (9H, s, 3 × CH$_3$), 1.89–1.94 (2H, m, CH$_2$), 3.63 (2H, t, J = 5.4 Hz, CH$_2$), 3.87 (2H, t, J = 6.0 Hz, CH$_2$), 5.24 (1H, d, J = 11.0 Hz, CH), 5.77 (1H, d, J = 17.6 Hz, CH), 6.69 (1H, dd, J = 17.6, 11.0 Hz, CH), 7.07 (1H, d,

$J = 8.3$ Hz, Ar), 7.40 (1H, dd, $J = 8.3$, 2.0 Hz, Ar), 8.20 (1H, d, $J = 2.0$ Hz, Ar); ^{13}C-NMR (100 MHz, CDCl$_3$) δ: 21.9, 30.0 (3C), 45.1, 45.4, 54.1, 114.1, 124.7, 126.6, 127.4, 127.8, 128.2, 135.7, 135.9, 138.2, 147.9; HRMS (FAB): m/z calcd for C$_{17}$H$_{22}$N$_3$S [M + H]$^+$ 300.1534; found: 300.1532.

Compound 26m. Using the general procedure as described for **25a**, compound **23m** (7.3 mg, 0.024 mmol) was allowed to react for 1 h with TFA (1.0 mL) and MS4Å (150 mg). Purification by flash chromatography over aluminum oxide with n-hexane–EtOAc (9:1) gave the title compound **26m** as colorless solid (3.7 mg, 62 %): mp 69–70 °C (from CHCl$_3$–n-hexane); IR (neat) cm^{-1}: 1622 (C=N), 1550 (C=N); ^{1}H-NMR (400 MHz, CDCl$_3$) δ: 1.95-2.01 (2H, m, CH$_2$), 3.70 (2H, t, $J = 5.7$ Hz, CH$_2$), 4.02 (2H, t, $J = 6.2$ Hz, CH$_2$), 5.27 (1H, dd, $J = 10.7$, 0.6 Hz, CH), 5.79 (1H, dd, $J = 17.7$, 0.6 Hz, CH), 6.69 (1H, dd, $J = 17.7$, 10.7 Hz, CH), 7.00 (1H, d, $J = 8.3$ Hz, Ar), 7.19 (1H, br s, NH), 7.42 (1H, dd, $J = 8.3$, 2.0 Hz, Ar), 8.23 (1H, d, $J = 2.0$ Hz, Ar); ^{13}C-NMR (100 MHz, CDCl$_3$) δ: 21.0, 43.8, 44.9, 114.5, 123.8, 126.8, 126.9, 127.8, 127.9, 135.6, 135.9, 146.5, 153.2; HRMS (FAB): m/z calcd for C$_{13}$H$_{14}$N$_3$S [M + H]$^+$ 244.0908; found: 244.0902.

3.1.31 Synthesis of 10-(4-Benzoylphenyl)-3,4-dihydro-2H, 6H-pyrimido[1,2-c][1,3]benzothiazin-6-imine (26q)

10-(4-Benzoylphenyl)-N-($tert$-butyl)-3,4-dihydro-2H,6H-pyrimido[1,2-c][1,3]benzothiazin-6-imine (23q). Using the general procedure as described for **22l**, 10-bromo-N-($tert$-butyl)-3,4-dihydro-2H,6H-pyrimido[1,2-c][1,3]benzothiazin-6-imine **23k** (52.8 mg, 0.15 mmol) was allowed to react with 4-benzoylphenylboronic acid (40.7 mg, 0.18 mmol) for 1 h. Purification by flash chromatography over aluminum oxide with n-hexane–EtOAc (8:2) gave the title compound **23q** as colorless solid (65.1 mg, 96 %): mp 192–193 °C (from CHCl$_3$–n-hexane); IR (neat) cm^{-1}: 1654 (C=O), 1592 (C=N); ^{1}H-NMR (400 MHz, CDCl$_3$) δ: 1.41 (9H, s, 3 × CH$_3$), 1.91–1.96 (2H, m, CH$_2$), 3.65 (2H, t, $J = 5.5$ Hz, CH$_2$), 3.90 (2H, t, $J = 6.1$ Hz, CH$_2$), 7.39 (1H, d, $J = 1.7$ Hz, Ar), 7.46–7.52 (3H, m, Ar), 7.58–7.62 (1H, m, Ar), 7.70 (2H, d, $J = 8.5$ Hz, Ar), 7.82 (2H, dd, $J = 8.3$, 1.2 Hz, Ar), 7.88 (2H, d, $J = 8.5$ Hz, Ar), 8.30 (1H, d, $J = 8.5$ Hz, Ar); ^{13}C-NMR (125 MHz, CDCl$_3$) δ: 21.9, 30.0 (3C), 45.1, 45.4, 54.2, 123.0, 124.8, 126.9 (2C), 127.3, 128.3 (2C), 129.1, 129.8, 130.0 (2C), 130.7 (2C), 132.5, 136.9, 137.6, 137.9, 141.6, 143.3, 147.5, 196.1; HRMS (FAB): m/z calcd for C$_{28}$H$_{28}$N$_3$OS [M + H]$^+$ 454.1953; found: 454.1952.

Compound 26q. Using the general procedure as described for **25a**, compound **23q** (36.2 mg, 0.08 mmol) was allowed to react for 1 h with TFA (1.0 mL) and MS4Å (150 mg). Purification by flash chromatography over aluminum oxide with n-hexane–EtOAc (7:3) gave the title compound **26q** as colorless solid (23.4 mg, 74 %): mp 163–165 °C (from CHCl$_3$–n-hexane); IR (neat) cm^{-1}: 1654 (C=O),

1622 (C=N), 1561 (C=N); ^{1}H-NMR (400 MHz, CDCl$_3$) δ: 1.97–2.03 (2H, m, CH$_2$), 3.72 (2H, t, J = 5.6 Hz, CH$_2$), 4.05 (2H, t, J = 6.2 Hz, CH$_2$), 7.15 (1H, d, J = 8.0 Hz, Ar), 7.48–7.52 (2H, m, Ar), 7.58–7.64 (2H, m, Ar), 7.73 (2H, d, J = 8.5 Hz, Ar), 7.82 (2H, dd, J = 8.2, 1.3 Hz, Ar), 7.88 (2H, d, J = 8.5 Hz, Ar), 8.57 (1H, d, J = 2.0 Hz, Ar); ^{13}C-NMR (100 MHz, CDCl$_3$) δ: 21.0, 43.9, 45.0, 124.3, 126.7 (2C), 127.2, 127.6, 128.3 (2C), 128.8, 129.2, 130.0 (2C), 130.7 (2C), 132.4, 136.5, 137.7, 138.0, 143.7, 146.3, 152.9, 196.1; HRMS (FAB): m/z calcd for C$_{24}$H$_{20}$N$_3$OS [M + H]$^+$ 398.1327; found: 398.1327.

3.1.32 Synthesis of 11-Fluoro-3,4-dihydro-2H, 6H-pyrimido[1,2-c][1,3]benzothiazin-6-imine (27i)

2-(2,6-Difluorophenyl)-1,4,5,6-tetrahydropyrimidine (20i). 2,6-Difluorobenzaldehyde **17i** (1.00 g, 7.04 mmol) was subjected to the general procedure as described for **18j** to give the title compound **20i** as colorless crystals (1.08 g, 78 %): mp 165–166 °C (from CHCl$_3$–n-hexane); IR (neat) cm^{-1}: 1620 (C=N); ^{1}H-NMR (500 MHz, CDCl$_3$) δ: 1.85–1.90 (2H, m, CH$_2$), 3.47 (4H, t, J = 5.7 Hz, 2 × CH$_2$), 4.77 (1H, br s, NH), 6.86–6.91 (2H, m, Ar), 7.24–7.30 (1H, m, Ar); ^{13}C-NMR (100 MHz, CDCl$_3$) δ: 20.5, 42.2 (2C), 111.4–111.6 (m, 2C), 115.9 (t, J = 20.3 Hz), 130.1 (t, J = 9.9 Hz), 146.8, 160.3 (dd, J = 250.3, 7.0 Hz, 2C). ^{19}F-NMR (500 MHz, CDCl$_3$) δ: −114.4; *Anal.* calcd for C$_{10}$H$_{10}$F$_2$N$_2$: C, 61.22; H, 5.14; N, 14.28. Found: C, 61.23; H, 5.13; N, 14.26.

N-(*tert*-Butyl)-11-fluoro-3,4-dihydro-2H,6H-pyrimido[1,2-c][1,3]benzothiazin-6-imine (24i). Using the general procedure as described for **22e**, compound **20i** (196.2 mg, 1.0 mmol) was allowed to react at rt overnight. Purification by flash chromatography over aluminum oxide with n-hexane–EtOAc (1:0 to 9:1) gave the title compound **24i** as colorless solid (212.6 mg, 73 %): mp 81 °C (from n-hexane); IR (neat) cm^{-1}: 1592 (C=N); ^{1}H-NMR (500 MHz, CDCl$_3$) δ: 1.37 (9H, s, 3 × CH$_3$), 1.90–1.95 (2H, m, CH$_2$), 3.66 (2H, t, J = 5.7 Hz, CH$_2$), 3.80 (2H, t, J = 6.6 Hz, CH$_2$), 6.94–6.99 (2H, m, Ar), 7.23–7.26 (1H, m, Ar); ^{13}C-NMR (125 MHz, CDCl$_3$) δ: 22.5, 30.1 (3C), 45.3, 45.5, 54.2, 115.0 (d, J = 24.0 Hz), 118.7, 120.9 (d, J = 3.6 Hz), 130.6 (d, J = 10.8 Hz), 131.9, 137.6, 146.1 (d, J = 8.4 Hz), 160.2 (d, J = 260.3 Hz); ^{19}F-NMR (500 MHz, CDCl$_3$) δ: −110.8; HRMS (FAB): m/z calcd for C$_{15}$H$_{19}$FN$_3$S [M + H]$^+$ 292.1284; found: 292.1288.

Compound 27i. Using the general procedure as described for 25a, compound **24i** (58.3 mg, 0.20 mmol) was allowed to react for 1 h. Purification by flash chromatography over aluminum oxide with n-hexane–EtOAc (9:1) gave the title compound **27i** as colorless solid (42.3 mg, 90 %): mp 142.5 °C (from CHCl$_3$–n-hexane); IR (neat) cm^{-1}: 1624 (C=N), 1585 (C=N); ^{1}H-NMR (500 MHz, CDCl$_3$) δ: 1.97–2.02 (2H, m, CH$_2$), 3.73 (2H, t, J = 5.2 Hz, CH$_2$), 3.94 (2H, t, J = 6.6 Hz, CH$_2$), 6.91 (1H, d, J = 8.0 Hz, Ar), 6.97–7.01 (1H, m, Ar), 7.22 (1H, br s, NH), 7.27–7.31 (1H, m, Ar); ^{13}C-NMR (125 MHz, CDCl$_3$) δ: 21.7, 44.0,

45.5, 115.3 (d, $J = 24.0$ Hz), 117.4 (d, $J = 8.4$ Hz), 120.0 (d, $J = 3.6$ Hz), 131.2 (d, $J = 9.6$ Hz), 131.5, 144.8 (d, $J = 9.6$ Hz), 152.6 (d, $J = 4.8$ Hz), 160.5 (d, $J = 261.5$ Hz); ^{19}F-NMR (500 MHz, CDCl$_3$) δ: -110.0. *Anal.* calcd for C$_{11}$H$_{10}$FN$_3$S: C, 56.15; H, 4.28; N, 17.86. Found: C, 56.05; H, 4.28; N, 17.71.

3.1.33 Synthesis of 8-Bromo-3,4-dihydro-2H, 6H-pyrimido[1,2-c][1,3]benzothiazin-6-imine (27k)

2-(3-Bromo-2-fluorophenyl)-1,4,5,6-tetrahydropyrimidine (20k). 3-Bromo-2-fluorobenzaldehyde 17k (0.71 g, 3.5 mmol) was subjected to the general procedure as described for **18j** to give the title compound **20k** as colorless crystals (0.62 g, 69 %): mp 99 °C; IR (neat) cm^{-1}: 1624 (C=N); ^{1}H-NMR (400 MHz, CDCl$_3$) δ: 1.84–1.89 (2H, m, CH$_2$), 3.50 (4H, t, $J = 5.7$ Hz, 2 × CH$_2$), 5.13 (1H, br s, NH), 7.03 (1H, td, $J = 8.0, 0.9$ Hz, Ar), 7.54 (1H, ddd, $J = 8.0, 6.4, 1.3$ Hz, Ar), 7.69 (1H, ddd, $J = 8.0, 6.5, 1.3$ Hz, Ar); ^{13}C-NMR (100 MHz, CDCl$_3$) δ: 20.6, 42.1 (2C), 109.6 (d, $J = 22.3$ Hz), 125.1 (d, $J = 4.1$ Hz), 126.3 (d, $J = 13.2$ Hz), 129.8 (d, $J = 3.3$ Hz), 134.3, 150.8, 156.3 (d, $J = 248.3$ Hz); ^{19}F-NMR (500 MHz, CDCl$_3$) δ: -110.7; *Anal.* calcd for C$_{10}$H$_{10}$BrFN$_2$: C, 46.72; H, 3.92; N, 10.90. Found: C, 46.64; H, 4.10; N, 10.93.

8-Bromo-*N*-(*tertH*,6*H*-pyrimido[1,2-c][1,3]benzothiazin-6-imine   (24k). Using the general procedure as described for **22e**, compound **20k** (257.1 mg, 1.00 mmol) was allowed to react at rt overnight. Purification by flash chromatography over aluminum oxide with *n*-hexane–EtOAc (1:0 to 9:1) gave the title compound **24k** as colorless solid (335.3 mg, 95 %): mp 89 °C (from *n*-hexane); IR (neat) cm^{-1}: 1595 (C=N); ^{1}H-NMR (500 MHz, CDCl$_3$) δ: 1.42 (9H, s, 3 × CH$_3$), 1.87–1.92 (2H, m, CH$_2$), 3.62 (2H, t, $J = 5.4$ Hz, CH$_2$), 3.86 (2H, t, $J = 6.0$ Hz, CH$_2$), 7.07 (1H, dd, $J = 8.0, 7.4$ Hz, Ar), 7.55 (1H, d, $J = 7.4$ Hz, Ar), 8.19 (1H, d, $J = 8.0$ Hz, Ar); ^{13}C-NMR (125 MHz, CDCl$_3$) δ: 21.7, 30.1 (3C), 45.3, 45.3, 54.3, 118.6, 126.5, 127.5, 129.9, 130.7, 133.8, 137.4, 147.5; *Anal.* calcd for C$_{15}$H$_{18}$BrN$_3$S: C, 51.14; H, 5.15; N, 11.93. Found: C, 50.89; H, 5.06; N, 11.83.

Compound 27k. Using the general procedure as described for **25a**, compound **24k** (52.8 mg, 0.15 mmol) was allowed to react for 2 h with TFA (1.5 mL) and MS4Å (225 mg). Purification by flash chromatography over silica gel with *n*-hexane–EtOAc (2:1) gave the title compound **27k** as colorless solid (31.6 mg, 71 %): mp 138–139 °C (from CHCl$_3$–*n*-hexane); IR (neat) cm^{-1}: 1567 (C=N); ^{1}H-NMR (500 MHz, CDCl$_3$) δ: 1.94–1.98 (2H, m, CH$_2$), 3.69 (2H, t, $J = 5.7$ Hz, CH$_2$), 4.02 (2H, t, $J = 6.0$ Hz, CH$_2$), 7.10 (1H, dd, $J = 8.3, 7.7$ Hz, Ar), 7.33 (1H, br s, NH), 7.56 (1H, dd, $J = 7.7, 1.4$ Hz, Ar), 8.23 (1H, dd, $J = 8.3, 1.4$ Hz, Ar); ^{13}C-NMR (125 MHz, CDCl$_3$) δ: 20.8, 43.7, 45.0, 117.6, 126.8, 127.9, 128.8, 130.5, 134.1, 146.2, 152.7; *Anal.* calcd for C$_{11}$H$_{10}$BrN$_3$S: C, 44.61; H, 3.40; N, 14.19. Found: C, 44.36; H, 3.64; N, 13.96.

3.1.34 Synthesis of 3,4-Dihydro-2H,6H-pyrimido[1,2-c]
naphtho[2,3-e][1,3]thiazin-6-imine (27r)

2-(3-Fluoronaphthalen-2-yl)-1,4,5,6-tetrahydropyrimidine (20r). 1-Fluoro-2-naphthaldehyde **17r** (0.96 g, 5.52 mmol) was subjected to the general procedure as described for **18j** to give the title compound **20r** as pale yellow crystals (0.85 g, 67 %): mp 128–130 °C (from CHCl$_3$–EtOAc–Et$_2$O); IR (neat) cm^{-1}: 1619 (C=N); ^{1}H-NMR (500 MHz, CDCl$_3$) δ: 1.86–1.91 (2H, m, CH$_2$), 3.52 (4H, t, J = 5.7 Hz, 2 × CH$_2$), 5.21 (1H, br s, NH), 7.41–7.44 (2H, m, Ar), 7.49 (1H, t, J = 7.4 Hz, Ar), 7.73 (1H, d, J = 8.6 Hz, Ar), 7.83 (1H, d, J = 8.6 Hz, Ar), 8.27 (1H, d, J = 8.0 Hz, Ar); ^{13}C-NMR (125 MHz, CDCl$_3$) δ: 20.5, 42.2 (2C), 111.7 (d, J = 22.8 Hz), 124.5 (d, J = 16.8 Hz), 125.6 (d, J = 2.4 Hz), 126.7 (d, J = 4.8 Hz), 127.5, 128.6, 130.1, 130.9 (d, J = 4.8 Hz), 134.2 (d, J = 9.6 Hz), 151.9, 157.8 (d, J = 247.1 Hz); ^{19}F-NMR (500 MHz, CDCl$_3$) δ: −119.8; HRMS (FAB) m/z calcd for C$_{14}$H$_{14}$FN$_2$ [M + H]$^+$ 229.1141; found: 229.1143.

N-(tert-Butyl)-3,4-dihydro-2H,6H-pyrimido[1,2-c]naphtho[2,3-e][1,3]thiazin-6-imine (24r). Using the general procedure as described for **22e**, compound **20r** (228.3 mg, 1.00 mmol) was allowed to react at rt overnight. Purification by flash chromatography over silica gel with n-hexane–EtOAc (1:1) gave the title compound **24r** as colorless solid (284.8 mg, 88 %): mp 82.5–83.5 °C, IR (neat) cm^{-1}: 1594 (C=N); ^{1}H-NMR (400 MHz, CDCl$_3$) δ: 1.42 (9H, s, 3 × CH$_3$), 1.92–1.98 (2H, m, CH$_2$), 3.69 (2H, t, J = 5.6 Hz, CH$_2$), 3.91 (2H, t, J = 6.2 Hz, CH$_2$), 7.38–7.43 (1H, m, Ar), 7.45–7.49 (1H, m, Ar), 7.60 (1H, s, Ar), 7.69 (1H, d, J = 7.8 Hz, Ar), 7.87 (1H, d, J = 8.0 Hz, Ar), 8.70 (1H, s, Ar); ^{13}C-NMR (100 MHz, CDCl$_3$) δ: 22.0, 30.1 (3C), 45.3, 45.5, 54.3, 122.5, 125.8, 125.9, 126.3, 126.5, 127.8, 128.5, 129.2, 131.7, 133.9, 138.4, 148.5; HRMS (FAB) m/z calcd for C$_{19}$H$_{22}$N$_3$S [M + H]$^+$ 324.1534; found: 324.1526.

Compound 27r. Using the general procedure as described for **25a**, compound **24r** (64.7 mg, 0.2 mmol) was allowed to react for 2 h. Purification by flash chromatography over aluminum oxide with n-hexane–EtOAc (4:1) gave the title compound **27r** as colorless solid (36.6 mg, 68 %): mp 180–181 °C (from CHCl$_3$–n-hexane); IR (neat) cm^{-1}: 1627 (C=N), 1572 (C=N); ^{1}H-NMR (400 MHz, CDCl$_3$) δ: 1.98–2.04 (2H, m, CH$_2$), 3.75 (2H, t, J = 5.5 Hz, CH$_2$), 4.06 (2H, t, J = 6.2 Hz, CH$_2$), 7.40–7.51 (3H, m, Ar), 7.68 (1H, d, J = 8.3 Hz, Ar), 7.87 (1H, d, J = 8.3 Hz, Ar), 8.74 (1H, s, Ar); ^{13}C-NMR (100 MHz, CDCl$_3$) δ: 21.1, 43.9, 45.1, 121.6, 125.0, 125.4, 126.1, 126.3, 128.1, 129.2, 129.2, 131.6, 133.9, 147.1, 153.4; HRMS (FAB) m/z calcd for C$_{15}$H$_{14}$N$_3$S [M + H]$^+$ 268.0908; found: 268.0909.

3.1.35 Synthesis of 3,4-Dihydro-2H,6H-pyrimido[1,2-c] pyrido[3,2-e][1,3]thiazin-6-imine (27s)

Using general procedure as described for **25f**, reaction of 2-(2-bromopyridin-3-yl)-1,4,5,6-tetrahydropyrimidine **21s** (58.8 mg, 0.25 mmol) and purification by flash chromatography over aluminum oxide with *n*-hexane–EtOAc (7:3) gave the title compound **27s** as colorless solid (17.4 mg, 32 %): mp 181–183 °C (from CHCl$_3$–*n*-hexane); IR (neat) cm^{-1}: 1624 (C=N), 1582 (C=N); ^{1}H-NMR (500 MHz, CDCl$_3$) δ: 1.96–2.01 (2H, m, CH$_2$), 3.70 (2H, t, J = 5.7 Hz, CH$_2$), 4.05 (2H, t, J = 6.3 Hz, CH$_2$), 7.17 (1H, dd, J = 8.0, 4.6 Hz, Ar), 7.39 (1H, br s, NH), 8.46–8.50 (2H, m, Ar); ^{13}C-NMR (125 MHz, CDCl$_3$) δ: 20.8, 43.8, 45.2, 121.4, 123.7, 136.3, 145.3, 151.2, 151.3, 153.5; HRMS (FAB): *m/z* calcd for C$_{10}$H$_{11}$N$_4$S [M + H]$^+$ 219.0704; found: 219.0703.

3.1.36 Synthesis of 2,3-Dihydronaphtho[2,1-e]pyrimido [1,2-c][1,3]thiazin-12(1H)- imine (27t)

2,3-Dihydronaphtho[2,1-*e*]pyrimido[1,2-*c*][1,3]thiazine-12(1H)-thione **21t** (71.1 mg, 0.25 mmol) was subjected to the general procedure as described for **25f** to give the title compound 27t as colorless solid (42.3 mg, 63 %): mp 157 °C (from CHCl$_3$–*n*-hexane); IR (neat) cm^{-1}: 1615 (C=N), 1572 (C=N); ^{1}H-NMR (500 MHz, CDCl$_3$) δ: 1.99–2.03 (2H, m, CH$_2$), 3.75 (2H, t, J = 5.4 Hz, CH$_2$), 4.07 (2H, t, J = 6.0 Hz, CH$_2$), 7.33 (1H, br s, NH), 7.53–7.57 (2H, m, Ar), 7.66 (1H, d, J = 8.6 Hz, Ar), 7.80–7.90 (2H, m, Ar), 8.30 (1H, d, J = 9.2 Hz, Ar); ^{13}C-NMR (125 MHz, CDCl$_3$) δ: 20.9, 43.7, 45.1, 123.2, 124.0, 125.1, 125.9, 126.7, 126.7, 127.5, 127.8, 128.5, 133.9, 147.1, 152.7; HRMS (FAB): *m/z* calcd for C$_{15}$H$_{14}$N$_3$S [M + H]$^+$ 268.0908; found: 268.0906.

3.1.37 Synthesis of 3,4-Dihydro-2H,6H-pyrimido[1,2-c] thieno[2,3-e][1,3]thiazin-6-imine (27u)

2-(3-Bromothiophen-2-yl)-1,4,5,6-tetrahydropyrimidine (20u). 3-Bromothiophene-2-carbaldehyde **17u** (1.29 g, 6.75 mmol) was subjected to the general procedure as described for **18j** to give the title compound **20u** as pale yellow crystals (1.11 g, 67 %): mp 61–63 °C (from CHCl$_3$–*n*-hexane); IR (neat) cm^{-1}: 1611 (C=N); ^{1}H-NMR (400 MHz, CDCl$_3$) δ: 1.83–1.88 (2H, m, CH$_2$), 3.48 (4H, t, J = 5.9 Hz, 2 × CH$_2$), 6.07 (1H, br s, NH), 6.92 (1H, d, J = 5.4 Hz, Ar), 7.24 (1H, d, J = 5.4 Hz, Ar); ^{13}C-NMR (100 MHz, CDCl$_3$) δ: 20.5, 42.0 (2C), 105.8, 127.3, 131.4, 135.6, 149.3; HRMS (FAB): *m/z* calcd for C$_8$H$_{10}$BrN$_2$S [M + H]$^+$ 244.9748; found: 244.9742.

3,4-Dihydro-2*H*,6*H*-pyrimido[1,2-*c*]thieno[2,3-*e*][1,3]thiazin-6-thione (21u).
To a mixture of **20u** (122.6 mg, 0.50 mmol) and NaH (40.0 mg, 1.0 mmol; 60 %
oil suspension) in DMF (1.7 mL) was added CS$_2$ (0.060 mL, 1.0 mmol) under an
Ar atmosphere. After being stirred at 80 °C for 12 h, the mixture was concen-
trated. The residue was purified by flash chromatography over silica gel with *n*-
hexane–EtOAc (8:2) to give the title compound **21u** as pale yellow solid (80.5 mg,
67 %): mp 167 °C (from CHCl$_3$–*n*-hexane); IR (neat) cm^{-1}: 1624 (C=N); ^{13}C-
NMR (400 MHz, CDCl$_3$) δ: 2.04–2.10 (2H, m, CH$_2$), 3.68 (2H, t, *J* = 5.5 Hz,
CH$_2$), 4.42 (2H, t, *J* = 6.1 Hz, CH$_2$), 6.76 (1H, d, *J* = 5.4 Hz, Ar), 7.49 (1H, d,
J = 5.4 Hz, Ar); ^{13}C-NMR (100 MHz, CDCl$_3$) δ: 21.5, 45.0, 48.5, 122.3, 128.4,
130.8, 131.0, 141.7, 189.7; HRMS (FAB): *m/z* calcd for C$_9$H$_9$N$_2$S$_3$ [M + H]$^+$
240.9928; found: 240.9936.

Compound 27u. Compound **21u** (60.1 mg, 0.25 mmol) was subjected to
general procedure as described for 25f to give the title compound **27u** as colorless
solid (19.4 mg, 35 %): mp 100–101 °C (from CHCl$_3$–*n*-hexane); IR (neat) cm^{-1}:
1616 (C=N), 1567 (C=N); ^{13}C-NMR (400 MHz, CDCl$_3$) δ: 1.99–2.05 (2H, m,
CH$_2$), 3.62 (2H, t, *J* = 5.6 Hz, CH$_2$), 3.99 (2H, t, *J* = 6.1 Hz, CH$_2$), 6.74 (1H, d,
J = 5.4 Hz, Ar), 7.28 (1H, br s, NH), 7.41 (1H, d, *J* = 5.4 Hz, Ar); ^{13}C-NMR
(100 MHz, CDCl$_3$) δ: 21.2, 43.5, 44.5, 123.3, 125.9, 127.0, 129.9, 143.7, 153.7;
Anal. calcd for C$_9$H$_9$N$_3$S$_2$: C, 48.40; H, 4.06; N, 18.82. Found: C, 48.38; H, 3.98;
N, 18.75.

3.1.38 Synthesis of 3,4-Dihydro-9-(4-methoxycarbonylphenyl)-2H,6H-pyrimido [1,2-c][1,3]benzothiazin-6-imine (31a)

**N-(*tert*-Butyl)-3,4-dihydro-9-(4-methoxycarbonylphenyl)-2*H*,6*H*-pyrimido [1,
2-*c*][1,3]benzothiazin-6-imine (29a).** Using the general procedure as described
for **22l**, N-(*tert*-butyl)-9-bromo-3,4-dihydro-2*H*,6*H*-pyrimido[1,2-*c*][1,3]benzo-
thiazin-6-imine **22k** (52.8 mg, 0.15 mmol) was allowed to react with 4-
(methoxycarbonyl)phenylboronic acid (32.4 mg, 0.18 mmol) for 1 h. Purification
by flash chromatography over aluminum oxide with *n*-hexane–EtOAc (1:0 to 9:1)
gave the title compound **29a** as colorless solid (47.3 mg, 77 %): mp 201–202 °C
(from CHCl$_3$–*n*-hexane); IR (neat) cm^{-1}: 1719 (C=O), 1593 (C=N); ^{1}H-NMR
(400 MHz, CDCl$_3$) δ: 1.40 (s, 9H, 3 × CH$_3$), 1.90–1.96 (m, 2H, CH$_2$), 3.65 (t,
J = 5.5 Hz, 2H, CH$_2$), 3.89 (t, *J* = 6.1 Hz, 2H, CH$_2$), 3.94 (s, 3H, CH$_3$), 7.36 (d,
J = 1.7 Hz, 1H, Ar), 7.44 (dd, *J* = 8.5, 1.7 Hz, 1H, Ar), 7.65 (d, *J* = 8.2 Hz, 2H,
Ar), 8.10 (d, *J* = 8.2 Hz, 2H, Ar), 8.28 (d, *J* = 8.5 Hz, 1H, Ar); ^{13}C-NMR
(100 MHz, CDCl$_3$) δ: 21.9, 30.0 (3C), 45.2, 45.4, 52.1, 54.2, 123.0, 124.8, 127.0
(2C), 127.3, 129.1, 129.6, 129.8, 130.2 (2C), 138.0, 141.7, 143.8, 147.5, 166.8;
HRMS (FAB): *m/z* calcd for C$_{23}$H$_{26}$N$_3$O$_2$S [M + H]$^+$ 408.1746; found: 408.1748.

Compound 31a. Using the general procedure as described for **25a**, compound **29a** (38.4 mg, 0.094 mmol) was allowed to react for 2 h. Purification by flash chromatography over aluminum oxide with n-hexane–EtOAc (9:1 to 1:1) gave the title compound **31a** as (27.3 mg, 83 %): mp 185–186 °C (from $CHCl_3$–n-hexane); IR (neat) cm^{-1}: 1719 (C=O), 1619 (C=N), 1566 (C=N); ^{1}H-NMR (500 MHz, $CDCl_3$) δ: 1.97–2.02 (m, 2H, CH_2), 3.71 (t, J = 5.7 Hz, 2H, CH_2), 3.94 (s, 3H, CH_3), 4.04 (t, J = 6.0 Hz, 2H, CH_2), 7.27 (d, J = 1.7 Hz, 1H, Ar), 7.46 (dd, J = 8.0, 1.7 Hz, 1H, Ar), 7.63 (d, J = 8.6 Hz, 2H, Ar), 8.10 (d, J = 8.6 Hz, 2H, Ar), 8.30 (d, J = 8.0 Hz, 1H, Ar); ^{13}C-NMR (125 MHz, $CDCl_3$) δ: 21.0, 43.8, 45.0, 52.2, 122.0, 125.1, 126.2, 126.9 (2C), 129.5, 129.6, 129.7, 130.2 (2C), 142.1, 143.4, 146.2, 153.0, 166.7; HRMS (FAB): m/z calcd for $C_{19}H_{18}N_3O_2S$ [M + H]$^+$ 352.1120; found: 352.1119.

3.1.39 Synthesis of 9-(4-Cyanophenyl)-3,4-dihydro-2H, 6H-pyrimido[1,2-c][1,3]benzothiazin-6-imine (31b)

N-(*tert*-Butyl)-9-(4-cyanophenyl)-3,4-dihydro-2*H*,6*H*-pyrimido[1,2-c][1,3]ben-zothiazin-6-imine (29b). Using the general procedure as described for **22l**, N -(*tert*-butyl)-9-bromo-3,4-dihydro-2*H*,6*H*-pyrimido[1,2-c][1,3]benzothiazin-6-imine **22k** (52.8 mg, 0.15 mmol) was allowed to react with 4-cyanophenylboronic acid (26.4 mg, 0.18 mmol) for 1 h. Purification by flash chromatography over aluminum oxide with n-hexane–EtOAc (8:2) gave the title compound **29b** as colorless solid (53.9 mg, 96 %): mp 188–190 °C (from $CHCl_3$–n-hexane); IR (neat) cm^{-1}: 2226 (C $\equiv$ N), 1593 (C=N); ^{1}H-NMR (400 MHz, $CDCl_3$) δ: 1.40 (s, 9H, 3 × CH_3), 1.91–1.96 (m, 2H, CH_2), 3.65 (t, J = 5.5 Hz, 2H, CH_2), 3.89 (t, J = 6.1 Hz, 2H, CH_2), 7.33 (d, J = 1.8 Hz, 1H, Ar), 7.41 (dd, J = 8.3, 1.8 Hz, 1H, Ar), 7.68 (d, J = 8.1 Hz, 2H, Ar), 7.73 (d, J = 8.1 Hz, 2H, Ar), 8.29 (d, J = 8.3 Hz, 1H, Ar); ^{13}C-NMR (100 MHz, $CDCl_3$) δ: 21.8, 30.0 (3C), 45.2, 45.4, 54.2, 111.7, 118.6, 123.0, 124.7, 127.6 (2C), 127.8, 129.3, 130.1, 132.6 (2C), 137.6, 140.7, 143.9, 147.4; HRMS (FAB): m/z calcd for $C_{22}H_{23}N_4S$ [M + H]$^+$ 375.1343; found: 375.1640.

Compound 31b. Using the general procedure as described for **25a**, compound **29b** (32.5 mg, 0.087 mmol) was allowed to react for 2 h. Purification by flash chromatography over aluminum oxide with n-hexane–EtOAc (6:4) gave the title compound **31b** as colorless solid (20.8 mg, 75 %): mp 210–211 °C (from $CHCl_3$–n-hexane); IR (neat) cm^{-1}: 2224 (C $\equiv$ N), 1615 (C=N), 1577 (C=N); ^{1}H-NMR (400 MHz, $CDCl_3$) δ: 1.97-2.02 (m, 2H, CH_2), 3.71 (d, J = 5.4 Hz, 2H, CH_2), 4.04 (t, J = 6.1 Hz, 2H, CH_2), 7.24 (d, J = 1.7 Hz, 1H, Ar), 7.43 (dd, J = 8.3, 1.7 Hz, 1H, Ar), 7.67 (d, J = 8.3 Hz, 2H, Ar), 7.73 (d, J = 8.3 Hz, 2H, Ar), 8.32 (d, J = 8.3 Hz, 1H, Ar); ^{13}C-NMR (100 MHz, $CDCl_3$) δ: 21.0, 43.8, 45.0, 111.9, 118.5, 122.0, 125.0, 126.7, 127.6 (2C), 129.7, 129.9, 132.7 (2C), 141.2, 143.5, 146.0, 152.8; HRMS (FAB): m/z calcd for $C_{18}H_{15}N_4S$ [M + H]$^+$ 319.1017; found: 319.1015.

3.1.40 Synthesis of 3,4-Dihydro-9-(4-nitrophenyl)-2H, 6H-pyrimido[1,2-c][1,3]benzothiazin-6-imine (31c)

N-(*tert*-Butyl)-3,4-dihydro-9-(4-nitrophenyl)-2*H*,6*H*-pyrimido[1,2-*c*][1,3]benzo thiazin-6-imine (29c). Using the general procedure as described for **22l**, N-(*tert*-butyl)-9-bromo-3,4-dihydro-2*H*,6*H*-pyrimido[1,2-*c*][1,3]benzothiazin-6-imine **22k** (352.3 mg, 1.0 mmol) was allowed to react with 4-nitrophenylboronic acid (200.3 mg, 1.2 mmol) for 1 h. Purification by flash chromatography over aluminum oxide with *n*-hexane–EtOAc (9:1 to 2:1) gave the title compound **29c** as colorless solid (366.9 mg, 93 %): mp 201–202 °C (from CHCl$_3$–*n*-hexane); IR (neat) cm^{-1}: 1590 (C=N), 1514 (NO$_2$); ^{1}H-NMR (400 MHz, CDCl$_3$) δ: 1.41 (s, 9H, 3 × CH$_3$), 1.91–1.97 (m, 2H, CH$_2$), 3.65 (t, J = 5.6 Hz, 2H, CH$_2$), 3.90 (t, J = 6.1 Hz, 2H, CH$_2$), 7.37 (d, J = 1.8 Hz, 1H, Ar), 7.44 (dd, J = 8.4, 1.8 Hz, 1H, Ar), 7.74 (d, J = 8.5 Hz, 2H, Ar), 8.28–8.32 (m, 3H, Ar); ^{13}C-NMR (100 MHz, CDCl$_3$) δ: 21.8, 30.0 (3C), 45.2, 45.5, 54.3, 123.2, 124.1 (2C), 124.8, 127.8 (2C), 128.0, 129.3, 130.2, 137.5, 140.4, 145.8, 147.4, 147.5; HRMS (FAB): *m/z* calcd for C$_{21}$H$_{23}$N$_4$O$_2$S [M + H]$^+$ 395.1542; found: 395.1539.

Compound 31c. Using the general procedure as described for **25a**, compound **29c** (77.0 mg, 0.075 mmol) was allowed to react for 4 h. Purification by flash chromatography over aluminum oxide with *n*-hexane–EtOAc (7:3) gave the title compound **31c** as colorless solid (50.7 mg, 75 %): mp 207–209 °C (from CHCl$_3$–*n*-hexane); IR (neat) cm^{-1}: 1615 (C=N), 1598 (NO$_2$), 1573 (C=N), 1508 (NO$_2$); ^{1}H-NMR (500 MHz, CDCl$_3$) δ: 1.98–2.02 (m, 2H, CH$_2$), 3.72 (t, J = 5.7 Hz, 2H, CH$_2$), 4.05 (t, J = 6.0 Hz, 2H, CH$_2$), 7.27 (s, 1H, Ar), 7.46 (d, J = 8.0 Hz, 1H, Ar), 7.72 (d, J = 8.6 Hz, 2H, Ar), 8.30 (d, J = 8.6 Hz, 2H, Ar), 8.34 (d, J = 8.0 Hz, 1H, Ar); ^{13}C-NMR (100 MHz, CDCl$_3$) δ: 21.0, 43.8, 45.0, 122.2, 124.2 (2C), 125.1, 126.9, 127.8 (2C), 129.7, 130.0, 140.8, 145.4, 146.0, 147.6, 152.7; HRMS (FAB): *m/z* calcd for C$_{17}$H$_{15}$N$_4$O$_2$S [M + H]$^+$ 339.0916; found: 339.0912.

3.1.41 Synthesis of 3,4-Dihydro-9-(4-trifluoromethylphenyl)-2H,6H-pyrimido[1,2-c][1,3]benzothiazin-6-imine (31d)

N-(*tert*-Butyl)-3,4-dihydro-9-(4-trifluoromethylphenyl)-2*H*,6*H*-pyrimido[1,2-*c*] [1,3]benzothiazin-6-imine (29d). Using the general procedure as described for **22l**, N-(*tert*-butyl)-9-bromo-3,4-dihydro-2*H*,6*H*-pyrimido[1,2-*c*][1,3]benzothiazin-6-imine **22k** (52.8 mg, 0.15 mmol) was allowed to react with 4-(trifluoro-methyl)phenylboronic acid (27.4 mg, 0.18 mmol) for 1 h. Purification by flash chromatography over aluminum oxide with *n*-hexane–EtOAc (1:0 to 9:1) gave the title compound **29d** as colorless solid (51.9 mg, 83 %): mp 177–179 °C (from *n*-hexane); IR (neat) cm^{-1}: 1594 (C=N); ^{1}H-NMR (400 MHz, CDCl$_3$) δ: 1.40 (s, 9H,

3 × CH$_3$), 1.90–1.96 (m, 2H, CH$_2$), 3.64 (t, J = 5.6 Hz, 2H, CH$_2$), 3.89 (t, J = 6.1 Hz, 2H, CH$_2$), 7.33 (d, J = 2.0 Hz, 1H, Ar), 7.41 (dd, J = 8.3, 2.0 Hz, 1H, Ar), 7.68 (s, 4H, Ar), 8.29 (d, J = 8.3 Hz, 1H, Ar); ^{13}C-NMR (125 MHz, CDCl$_3$) δ: 21.8, 29.9 (3C), 45.1, 45.4, 54.2, 122.9, 124.1 (q, J = 271.1 Hz), 124.8, 125.8 (q, J = 3.6 Hz, 2C), 127.3 (2C), 127.4, 129.2, 129.9, 130.0 (q, J = 32.8 Hz), 137.8, 141.3, 142.9, 147.5; ^{19}F-NMR (500 MHz, CDCl$_3$) δ: −63.0; HRMS (FAB): m/z calcd for C$_{22}$H$_{23}$F$_3$N$_3$S [M + H]$^+$ 418.1565; found: 418.1563.

Compound 31d. Using the general procedure as described for **25a**, compound **29d** (41.2 mg, 1.0 mmol) was allowed to react for 4 h with TFA (1.0 mL) and MS4Å (150 mg). Purification by flash chromatography over aluminum oxide with *n*-hexane–EtOAc (7:3) gave the title compound **31d** as colorless solid (26.0 mg, 73 %): mp 142–143 °C (from CHCl$_3$–*n*-hexane); IR (neat) cm^{-1}: 1619 (C=N), 1567 (C=N); ^{1}H-NMR (500 MHz, CDCl$_3$) δ: 1.95–2.00 (m, 2H, CH$_2$), 3.70 (t, J = 5.4 Hz, 2H, CH$_2$), 4.03 (t, J = 6.0 Hz, 2H, CH$_2$), 7.22 (s, 1H, Ar), 7.42 (d, J = 8.0 Hz, 1H, Ar), 7.64 (d, J = 8.0 Hz, 2H, Ar), 7.68 (d, J = 8.0 Hz, 2H, Ar), 8.30 (d, J = 8.0 Hz, 1H, Ar); ^{13}C-NMR (125 MHz, CDCl$_3$) δ: 20.9, 43.8, 44.9, 121.9, 124.0 (q, J = 272.2 Hz, 125.0, 125.7 (t, J = 3.6 Hz, 2C), 126.2, 127.2 (2C), 129.5, 129.6, 130.1 (q, J = 32.4 Hz), 141.7, 142.5, 146.1, 152.8; ^{19}F-NMR (500 MHz, CDCl$_3$) δ: −63.1; *Anal.* calcd for C$_{18}$H$_{14}$F$_3$N$_3$S: C, 59.82; H, 3.90; N, 11.63. Found: C, 59.56; H, 3.81; N, 11.48.

3.1.42 Synthesis of 9-(4-Aminocarbonylphenyl)-3,4-dihydro-2H,6H-pyrimido[1,2-c][1,3]benzothiazin-6-imine (31e)

9-[(4-Aminocarbonyl)phenyl]-*N*-(*tert*-butyl)-3,4-dihydro-2H,6H-pyrimido[1,2-c][1,3]benzothiazin-6-imine (29e). Using the general procedure as described for **22l**, *N*-(*tert*-butyl)-9-bromo-3,4-dihydro-2H,6H-pyrimido[1,2-c][1,3]benzothiazin-6-imine **22k** (52.8 mg, 0.15 mmol) was allowed to react with 4-(aminocarbonyl)phenylboronic acid (29.7 mg, 0.18 mmol) in 1,4-dioxane for 1 h. Purification by preparative TLC over aluminum oxide with CHCl$_3$–MeOH (95:5) gave the title compound **29e** as colorless solid (31.2 mg, 53 %): mp 261–263 °C (from MeOH–CHCl$_3$–*n*-hexane); IR (neat) cm^{-1}: 1650 (C=O), 1592 (C=N); ^{1}H-NMR (400 MHz, CDCl$_3$–CD$_3$OD) δ: 1.41 (s, 9H, 3 × CH$_3$), 1.92–1.98 (m, 2H, CH$_2$), 3.62 (t, J = 5.5 Hz, 2H, CH$_2$), 3.90 (t, J = 6.1 Hz, 2H, CH$_2$), 7.38 (d, J = 1.7 Hz, 1H, Ar), 7.47 (dd, J = 8.5, 1.7 Hz, 1H, Ar), 7.66 (d, J = 8.3 Hz, 2H, Ar), 7.92 (d, J = 8.3 Hz, 2H, Ar), 8.17 (d, J = 8.5 Hz, 1H, Ar); ^{13}C-NMR (100 MHz, CDCl$_3$–CD$_3$OD) δ: 21.6, 29.7 (3C), 44.7, 45.4, 54.2, 122.9, 124.8, 126.8, 126.9 (2C), 128.0 (2C), 128.8, 129.8, 132.6, 137.8, 141.8, 142.6, 148.7, 169.7; HRMS (FAB): m/z calcd for C$_{22}$H$_{25}$N$_4$OS [M + H]$^+$ 393.1749; found: 393.1744.

Compound 31e. TFA (17 mL) was added to a mixture of **29e** (27.1 mg, 0.069 mmol) and MS4 Å (4.5 g, powder, activated by heating with Bunsen burner) in CHCl$_3$ (3.0 mL) and MeOH (10 drops). After being stirred under reflux for 9 h, the mixture was concentrated. To a stirring mixture of this residue in CHCl$_3$ was added dropwise Et$_3$N at 0 °C to adjust pH to 8–9. The whole was extracted with EtOAc. The extract was washed with sat. NaHCO$_3$, brine, and dried over MgSO$_4$. After concentration, the residue was purified by preparative TLC over aluminum oxide with EtOAc–MeOH (95:5) to give compound **31e** as colorless solid (13.0 mg, 56 %): mp 222–223 °C (from MeOH–CHCl$_3$–n-hexane); IR (neat) cm^{-1}: 1666 (C=O), 1616 (C=N), 1556 (C=N); ^{1}H-NMR (400 MHz, DMSO-d_6) δ: 1.83–1.89 (m, 2H, CH$_2$), 3.59 (t, J = 5.1 Hz, 2H, CH$_2$), 3.92 (t, J = 5.7 Hz, 2H, CH$_2$), 7.40 (s, 1H, NH), 7.60–7.62 (m, 2H, Ar), 7.81 (d, J = 8.3 Hz, 2H, Ar), 7.96 (d, J = 8.3 Hz, 2H, Ar), 8.04 (s, 1H, NH), 8.24 (d, J = 8.3 Hz, 1H, Ar), 8.74 (s, 1H, NH); ^{13}C-NMR (100 MHz, DMSO-d_6) δ: 20.7, 43.1, 44.4, 121.8, 124.3, 125.4, 126.5 (2C), 128.1 (2C), 129.0, 129.7, 133.8, 140.6, 141.1, 145.1, 149.5, 167.3; HRMS (FAB): m/z calcd for C$_{18}$H$_{17}$N$_4$OS [M + H]$^+$ 337.1123; found: 337.1118.

3.1.43 Synthesis of 3,4-Dihydro-9-(4-methoxyphenyl)- 2H,6H-pyrimido[1,2-c][1,3]benzothiazin-6-imine (31f)

N-(tert-Butyl)-3,4-dihydro-9-(4-methoxyphenyl)-2H,6H-pyrimido[1,2-c][1,3] benzothiazin-6-imine (29f). Using the general procedure as described for **22l**, N-(tert-butyl)-9-bromo-3,4-dihydro-2H,6H-pyrimido[1,2-c][1,3]benzothiazin-6-imine **22k** (52.8 mg, 0.15 mmol) was allowed to react with 4-methoxyphenylboronic acid (27.4 mg, 0.18 mmol) for 1 h. Purification by flash chromatography over aluminum oxide with n-hexane–EtOAc (1:0 to 9:1) gave the title compound **29f** as colorless solid (54.7 mg, 96 %): mp 199–200 °C (from CHCl$_3$–n-hexane); IR (neat) cm^{-1}: 1592 (C=N); ^{1}H-NMR (400 MHz, CDCl$_3$) δ: 1.40 (s, 9H, 3 × CH$_3$), 1.89–1.95 (m, 2H, CH$_2$), 3.63 (t, J = 5.5 Hz, 2H, CH$_2$), 3.84 (s, 3H, CH$_3$), 3.88 (t, J = 6.0 Hz, 2H, CH$_2$), 6.96 (d, J = 8.5 Hz, 2H, Ar), 7.29 (d, J = 2.0 Hz, 1H, Ar), 7.38 (dd, J = 8.3, 2.0 Hz, 1H, Ar), 7.53 (d, J = 8.5 Hz, 2H, Ar), 8.22 (d, J = 8.3 Hz, 1H, Ar); ^{13}C-NMR (100 MHz, CDCl$_3$) δ: 21.9, 30.0 (3C), 45.1, 45.4, 54.1, 55.3, 114.3 (2C), 122.1, 124.4, 125.9, 128.1 (2C), 128.9, 129.4, 131.9, 138.4, 142.5, 147.8, 159.8; HRMS (FAB): m/z calcd for C$_{22}$H$_{26}$N$_3$OS [M + H]$^+$ 380.1797; found: 380.1801.

Compound 31f. Using the general procedure as described for **25a**, compound **29f** (28.4 mg, 0.075 mmol) was allowed to react for 2 h with TFA (1.0 mL) and MS4Å (150 mg). Purification by flash chromatography over aluminum oxide with n-hexane–EtOAc (3:1) gave the title compound **31f** as colorless solid (20.0 mg, 82 %): mp 93–94 °C (from CHCl$_3$–n-hexane); IR (neat) cm^{-1}: 1620 (C=N), 1567 (C=N); ^{1}H-NMR (400 MHz, CDCl$_3$) δ: 1.96-2.02 (m, 2H, CH$_2$), 3.71 (t, J = 5.5 Hz, 2H, CH$_2$), 3.85 (s, 3H, CH$_3$), 4.04 (t, J = 6.2 Hz, 2H, CH$_2$), 6.97 (d, J = 8.5 Hz, 2H, Ar), 7.20 (d, J = 1.7 Hz, 1H, Ar), 7.41 (dd, J = 8.3, 1.7 Hz, 1H,

Ar), 7.52 (d, $J = 8.5$ Hz, 2H, Ar), 8.25 (d, $J = 8.3$ Hz, 1H, Ar); ^{13}C-NMR (100 MHz, CDCl$_3$) δ: 21.0, 43.8, 44.9, 55.4, 114.4 (2C), 121.1, 124.7, 124.8, 128.1 (2C), 129.3, 129.3, 131.5, 143.1, 146.5, 153.5, 159.9; *Anal.* calcd for C$_{18}$H$_{17}$N$_3$OS: C, 66.85; H, 5.30; N, 12.99. Found: C, 66.95; H, 5.50; N, 12.89.

3.1.44 Synthesis of 3,4-Dihydro-9-(4-methylthiophenyl)-2H,6H-pyrimido[1,2-c][1,3]benzothiazin-6-imine (31g)

N-(*tert*-**Butyl**)-**3,4-dihydro-9-(4-methylthiophenyl)-2***H*,6*H*-**pyrimido[1,2-***c***][1, 3]benzothiazin-6-imine (29g).** Using the general procedure as described for **22l**, *N*-(*tert*-butyl)-9-bromo-3,4-dihydro-2*H*,6*H*-pyrimido[1,2-*c*][1,3]benzothiazin-6-imine **22k** (52.8 mg, 0.15 mmol) was allowed to react with 4-(methylthio)phenylboronic acid (30.2 mg, 0.18 mmol) for 1 h. Purification by flash chromatography over aluminum oxide with *n*-hexane–EtOAc (1:0 to 9:1) gave the title compound **29g** as colorless solid (50.6 mg, 85 %): mp 201–202 °C (from CHCl$_3$–*n*-hexane); IR (neat) cm^{-1}: 1592 (C=N); ^{1}H-NMR (400 MHz, CDCl$_3$) δ: 1.40 (s, 9H, 3 × CH$_3$), 1.89–1.95 (m, 2H, CH$_2$), 2.51 (s, 3H, CH$_3$), 3.64 (t, $J = 5.5$ Hz, 2H, CH$_2$), 3.88 (t, $J = 6.1$ Hz, 2H, CH$_2$), 7.30–7.32 (m, 3H, Ar), 7.40 (dd, $J = 8.4$, 1.3 Hz, 1H, Ar), 7.51 (d, $J = 8.3$ Hz, 2H, Ar), 8.23 (d, $J = 8.4$ Hz, 1H, Ar; ^{13}C-NMR (100 MHz, CDCl$_3$) δ: 15.7, 21.9, 30.0 (3C), 45.1, 45.4, 54.2, 122.3, 124.4, 126.4, 126.8 (2C), 127.3 (2C), 129.0, 129.6, 136.0, 138.2, 138.8, 142.2, 147.7; HRMS (FAB): *m/z* calcd for C$_{22}$H$_{26}$N$_3$S$_2$ [M + H]$^+$ 396.1568; found: 396.1566.

    Compound 31g. Using the general procedure as described for **25a**, compound **29g** (38.5 mg, 0.097 mmol) was allowed to react for 4 h. Purification by flash chromatography over aluminum oxide with *n*-hexane–EtOAc (9:1 to 7:3) gave the title compound **31g** as colorless solid (18.4 mg, 56 %): mp 151–153 °C (from CHCl$_3$–*n*-hexane); IR (neat) cm^{-1}: 1620 (C=N), 1573 (C=N); ^{1}H-NMR (400 MHz, CDCl$_3$) δ: 1.96–2.02 (m, 2H, CH$_2$), 2.52 (s, 3H, CH$_3$), 3.71 (t, $J = 5.5$ Hz, 2H, CH$_2$), 4.04 (t, $J = 6.1$ Hz, 2H, CH$_2$), 7.21 (d, $J = 1.8$ Hz, 1H, Ar), 7.31 (d, $J = 8.3$ Hz, 2H, Ar), 7.42 (dd, $J = 8.5$, 1.8 Hz, 1H, Ar), 7.49 (d, $J = 8.3$ Hz, 2H, Ar), 8.27 (d, $J = 8.5$ Hz, 1H, Ar); ^{13}C-NMR (100 MHz, CDCl$_3$) δ: 15.6, 21.0, 43.8, 44.9, 121.3, 124.7, 125.3, 126.7 (2C), 127.3 (2C), 129.4 (2C), 135.7, 139.2, 142.8, 146.5, 153.3; HRMS (FAB): *m/z* calcd for C$_{18}$H$_{18}$N$_3$S$_2$ [M + H]$^+$ 340.0942; found: 340.0944.

3.1.45 Synthesis of 3,4-Dihydro-9-(4-trifluomethoxyphenyl)-2H,6H-pyrimido[1,2-c][1,3]benzothiazin-6-imine (31h)

N-(tert-Butyl)-3,4-dihydro-9-(4-trifluoromethoxyphenyl)-2H,6H-pyrimido [1, 2-c][1,3]benzothiazin-6-imine (29h). Using the general procedure as described for **22l**, N-(tert-butyl)-9-bromo-3,4-dihydro-2H,6H-pyrimido[1,2-c][1,3]benzo-thiazin-6-imine **22k** (52.8 mg, 0.15 mmol) was allowed to react with 4-(trifluoromethoxy)phenylboronic acid (37.1 mg, 0.18 mmol) for 1 h. Purification by flash chromatography over aluminum oxide with n-hexane–EtOAc (1:0 to 9:1) gave the title compound **29h** as colorless solid (59.7 mg, 92 %): mp 157 °C (from CHCl$_3$–n-hexane); IR (neat) cm^{-1}: 1595 (C=N); ^{1}H-NMR (500 MHz, CDCl$_3$) δ: 1.40 (s, 9H, 3 × CH$_3$), 1.91-1.95 (m, 2H, CH$_2$), 3.64 (t, J = 5.4 Hz, 2H, CH$_2$), 3.89 (t, J = 6.0 Hz, 2H, CH$_2$), 7.27–7.30 (m, 3H, Ar), 7.38 (dd, J = 8.0, 1.7 Hz, 1H, Ar), 7.59 (d, J = 8.6 Hz, 2H, Ar), 8.26 (d, J = 8.0 Hz, 1H, Ar); ^{13}C-NMR (100 MHz, CDCl$_3$) δ: 21.9, 30.0 (3C), 45.2, 45.5, 54.2, 120.5 (q, J = 257.4 Hz), 121.3 (2C), 122.8, 124.7, 127.0, 128.5 (2C), 129.1, 129.8, 138.0, 138.2, 141.5, 147.6, 149.2 (q, J = 1.7 Hz); ^{19}F-NMR (500 MHz, CDCl$_3$) δ: −58.3; HRMS (FAB): m/z calcd for C$_{22}$H$_{23}$F$_3$N$_3$OS [M + H]$^+$ 434.1514; found: 434.1512.

Compound 31h. Using the general procedure as described for **25a**, compound **29h** (44.8 mg, 0.103 mmol) was allowed to react for 2 h. Purification by flash chromatography over aluminum oxide with n-hexane–EtOAc (9:1–7:3) gave the title compound **31h** as colorless solid (17.3 mg, 45 %): mp 120 °C (from CHCl$_3$–n-hexane); IR (neat) cm^{-1}: 1621 (C=N), 1571 (C=N); ^{1}H-NMR (500 MHz, CDCl$_3$) δ: 1.98-2.02 (m, 2H, CH$_2$), 3.72 (t, J = 5.7 Hz, 2H, CH$_2$), 4.04 (t, J = 6.3 Hz, 2H, CH$_2$), 7.22 (d, J = 1.7 Hz, 1H, Ar), 7.29 (d, J = 8.6 Hz, 2H, Ar), 7.41 (dd, J = 8.0, 1.7 Hz, 1H, Ar), 7.59 (d, J = 8.6 Hz, 2H, Ar), 8.29 (d, J = 8.6 Hz, 1H, Ar); ^{13}C-NMR (125 MHz, CDCl$_3$) δ: 20.9, 43.8, 44.9, 120.4 (q, J = 257.5 Hz), 121.3 (2C), 121.7, 124.9, 125.8, 128.4 (2C), 129.4, 129.5, 137.8, 141.9, 146.2, 149.2, 153.0; ^{19}F-NMR (500 MHz, CDCl$_3$) δ: −58.4; HRMS (FAB): m/z calcd for C$_{18}$H$_{15}$F$_3$N$_3$OS [M + H]$^+$ 378.0888; found: 378.0888.

3.1.46 Synthesis of 3,4-Dihydro-9-(3-methoxycarbonylphenyl)-2H,6H-pyrimido[1,2-c][1,3]benzothiazin-6-imine (31i)

N-(-(tert-Butyl)-3,4-dihydro-9-(3-methoxycarbonylphenyl)-2H,6Hc][1,3]benzo thiazin-6-imine (29i). Using the general procedure as described for **22l**, N-(tert-butyl)-9-bromo-3,4-dihydro-2H,6H-pyrimido[1,2-c][1,3]benzothiazin-6-imine **22k** (52.8 mg, 0.15 mmol) was allowed to react with 3-(methoxycarbonyl)phenylboronic acid (32.4 mg, 0.18 mmol) for 1 h. Purification by flash chromatography over aluminum oxide with *n*-hexane–EtOAc (1:0 to 9:1) gave the title compound **29i** as colorless solid (56.2 mg, 92 %): mp 116–117.5 °C (from *n*-hexane); IR (neat) cm^{-1}: 1723 (C=O), 1592 (C=N); ^{1}H-NMR (500 MHz, CDCl$_3$) δ: 1.41 (s, 9H, 3 × CH$_3$), 1.91-1.96 (m, 2H, CH$_2$), 3.65 (t, J = 5.4 Hz, 2H, CH$_2$), 3.89 (t, J = 6.0 Hz, 2H, CH$_2$), 3.95 (s, 3H, CH$_3$), 7.37 (d, J = 1.7 Hz, 1H, Ar), 7.45 (dd, J = 8.6, 1.7 Hz, 1H, Ar), 7.51 (t, J = 8.0 Hz, 1H, Ar), 7.78 (d, J = 8.0 Hz, 1H, Ar), 8.04 (d, J = 8.0 Hz, 1H, Ar), 8.26-8.28 (m, 2H, Ar); ^{13}C-NMR (125 MHz, CDCl$_3$) δ: 21.9, 30.0 (3C), 45.1, 45.4, 52.2, 54.2, 122.8, 124.7, 127.0, 128.0, 129.0, 129.1 (2C), 129.7, 130.8, 131.3, 138.1, 139.7, 141.7, 147.6, 166.8; HRMS (FAB): *m/z* calcd for C$_{23}$H$_{26}$N$_3$O$_2$S [M + H]$^+$ 408.1746; found: 408.1741.

Compound 31i. Using the general procedure as described for **25a**, compound **29i** (34.2 mg, 0.084 mmol) was allowed to react for 3 h. Purification by flash chromatography over aluminum oxide with *n*-hexane–EtOAc (7:3) gave the title compound **31i** as colorless solid (22.8 mg, 77 %): mp 131 °C (from CHCl$_3$–*n*-hexane); IR (neat) cm^{-1}: 1721 (C=O), 1620 (C=N), 1568 (C=N); ^{1}H-NMR (500 MHz, CDCl$_3$) δ: 1.97-2.02 (m, 2H, CH$_2$), 3.72 (t, J = 5.4 Hz, 2H, CH$_2$), 3.96 (s, 3H, CH$_3$), 4.04 (t, J = 6.0 Hz, 2H, CH$_2$), 7.28 (d, J = 1.4 Hz, 1H, Ar), 7.47 (dd, J = 8.6, 1.4 Hz, 1H, Ar), 7.52 (t, J = 7.7 Hz, 1H, Ar), 7.76 (d, J = 7.7 Hz, 1H, Ar), 8.05 (d, J = 7.7 Hz, 1H, Ar), 8.25 (s, 1H, Ar), 8.30 (d, J = 8.6 Hz, 1H, Ar); ^{13}C-NMR (125 MHz, CDCl$_3$) δ: 21.0, 43.8, 45.0, 52.3, 121.8, 125.0, 125.9, 128.1, 129.0, 129.2, 129.5, 129.5, 130.9, 131.3, 139.4, 142.3, 146.3, 153.1, 166.7; *Anal.* calcd for C$_{19}$H$_{17}$N$_3$O$_2$S: C, 64.94; H, 4.88; N, 11.96. Found: C, 64.83; H, 4.79; N, 11.84.

3.1.47 Synthesis of 9-(3-Cyanophenyl)-3,4-dihydro-2H, 6H-pyrimido[1,2-c][1,3]benzothiazin-6-imine (31j)

N-(*tert*-Butyl)-9-(3-cyanophenyl)-3,4-dihydro-2H,6H-pyrimido[1,2-c][1,3]ben-zothiazin-6-imine (29j). Using the general procedure as described for **22l**, N-(*tert*-butyl)-9-bromo-3,4-dihydro-2H,6H-pyrimido[1,2-c][1,3]benzothiazin-6-imine **22k** (52.8 mg, 0.15 mmol) was allowed to react with 3-cyanophenylboronic acid (26.5 mg, 0.18 mmol) for 1 h. Purification by flash chromatography over aluminum oxide with *n*-hexane–EtOAc (85:5) gave the title compound **29j** as colorless solid (48.4 mg, 86 %): mp 165–167 °C (from CHCl$_3$–*n*-hexane); IR (neat) cm^{-1}: 2230 (C ≡ N), 1593 (C=N); ^{1}H-NMR (500 MHz, CDCl$_3$) δ: 1.40 (s, 9H, 3 × CH$_3$), 1.91-1.96 (m, 2H, CH$_2$), 3.65 (t, J = 5.7 Hz, 2H, CH$_2$), 3.89 (t, J = 6.3 Hz, 2H, CH$_2$), 7.31 (d, J = 1.7 Hz, 1H, Ar), 7.38 (dd, J = 8.0, 1.7 Hz, 1H, Ar), 7.55 (t, J = 7.7 Hz, 1H, Ar), 7.65 (dt, J = 7.7, 1.7 Hz, 1H, Ar), 7.81 (dt, J = 7.7, 1.7 Hz, 1H, Ar), 7.86 (t, J = 1.7 Hz, 1H, Ar), 8.29 (d, J = 8.0 Hz, 1H, Ar); ^{13}C-NMR (100 MHz, CDCl$_3$) δ: 21.8, 30.0 (3C), 45.1, 45.4, 54.2, 113.2, 118.5, 122.8, 124.5,

127.6, 129.3, 129.7, 130.1, 130.5, 131.3, 131.4, 137.7, 140.4, 140.8, 147.4; HRMS (FAB): m/z calcd for $C_{22}H_{23}N_4S$ [M + H]$^+$ 375.1643; found: 375.1646.

Compound 31j. Using the general procedure as described for **25a**, compound **29j** (40.3 mg, 0.11 mmol) was allowed to react for 2 h. Purification by flash chromatography over aluminum oxide with n-hexane–EtOAc (2:1 to 0:1) gave the title compound **31j** as colorless solid (30.6 mg, 89 %): mp 196–197 °C (from MeOH–CHCl$_3$–n-hexane); IR (neat) cm^{-1}: 2230 (C $\equiv$ N), 1623 (C=N), 1572 (C=N); ^{1}H-NMR (400 MHz, CDCl$_3$–CD$_3$OD) δ: 1.98–2.04 (m, 2H, CH$_2$), 3.71 (t, J = 5.5 Hz, 2H, CH$_2$), 4.01 (t, J = 6.1 Hz, 2H, CH$_2$), 7.25 (d, J = 1.8 Hz, 1H, Ar), 7.44 (dd, J = 8.5, 1.8 Hz, 1H, Ar), 7.58 (t, J = 7.8 Hz, 1H, Ar), 7.68 (d, J = 7.8 Hz, 1H, Ar), 7.81 (d, J = 7.8 Hz, 1H, Ar), 7.86 (s, 1H, Ar), 8.27 (d, J = 8.5 Hz, 1H, Ar); ^{13}C-NMR (100 MHz, CDCl$_3$–CD$_3$OD) δ: 20.7, 43.9, 44.7, 113.0, 118.3, 121.9, 124.9, 126.2, 129.5, 129.7, 129.7, 130.4, 131.3, 131.5, 140.2, 141.0, 146.6, 153.4; HRMS (FAB): m/z calcd for $C_{18}H_{15}N_4S$ [M + H]$^+$ 319.1017; found: 319.1016.

3.1.48 Synthesis of 3,4-Dihydro-9-(3-nitrophenyl)-2H,6H-pyrimido[1,2-c][1,3]benzothiazin-6-imine (31k)

N-(tert-Butyl)-3,4-dihydro-9-(3-nitrophenyl)-2H,6H-pyrimido[1,2-c][1,3]benzothiazin-6-imine (29k). Using the general procedure as described for **22l**, N-(*tert*-butyl)-9-bromo-3,4-dihydro-2H,6H-pyrimido[1,2-c][1,3]benzothiazin-6-imine **22k** (52.8 mg, 0.15 mmol) was allowed to react with 3-nitrophenylboronic acid (30.0 mg, 0.18 mmol) for 1 h. Purification by flash chromatography over aluminum oxide with n-hexane–EtOAc (1:0 to 8:2) gave the title compound **29k** as colorless solid (52.3 mg, 88 %): mp 208–209 °C (from CHCl$_3$–n-hexane); IR (neat) cm^{-1}: 1593 (C=N), 1529 (NO$_2$); ^{1}H-NMR (500 MHz, CDCl$_3$) δ: 1.41 (s, 9H, 3 × CH$_3$), 1.92–1.96 (m, 2H, CH$_2$), 3.65 (t, J = 5.4 Hz, 2H, CH$_2$), 3.90 (t, J = 6.3 Hz, 2H, CH$_2$), 7.37 (d, J = 1.7 Hz, 1H, Ar), 7.44 (dd, J = 8.6, 1.7 Hz, 1H, Ar), 7.62 (t, J = 8.0 Hz, 1H, Ar), 7.91 (dd, J = 8.0, 2.0 Hz, 1H, Ar), 8.22 (dd, J = 8.0, 2.0 Hz, 1H, Ar), 8.31 (d, J = 8.6 Hz, 1H, Ar), 8.44 (t, J = 2.0 Hz, 1H, Ar); ^{13}C-NMR (125 MHz, CDCl$_3$) δ: 21.8, 30.0 (3C), 45.2, 45.4, 54.3, 121.8, 122.7, 122.9, 124.6, 127.7, 129.4, 129.9, 130.2, 132.9, 137.6, 140.2, 141.1, 147.4, 148.7; HRMS (FAB): m/z calcd for $C_{21}H_{23}N_4O_2S$ [M + H]$^+$ 395.1542; found: 395.1544.

Compound 31k. Using the general procedure as described for **25a**, compound **29k** (37.9 mg, 0.096 mmol) was allowed to react for 3 h. Purification by flash chromatography over aluminum oxide with n-hexane–EtOAc (1:1) gave the title compound **31k** as colorless solid (23.2 mg, 71 %): mp 168–170 °C (from CHCl$_3$–n-hexane); IR (neat) cm^{-1}: 1620 (C=N), 1567 (C=N), 1529 (NO$_2$); ^{1}H-NMR (500 MHz, CDCl$_3$) δ: 1.98-2.03 (m, 2H, CH$_2$), 3.72 (t, J = 5.7 Hz, 2H, CH$_2$), 4.05 (t, J = 6.0 Hz, 2H, CH$_2$), 7.28 (d, J = 1.7 Hz, 1H, Ar), 7.47 (dd, J = 8.6, 1.7 Hz,

1H, Ar), 7.63 (t, $J = 7.7$ Hz, 1H, Ar), 7.90 (dd, $J = 7.7$, 1.7 Hz, 1H, Ar), 8.23 (dd, $J = 7.7$, 1.7 Hz, 1H, Ar), 8.34 (d, $J = 8.6$ Hz, 1H, Ar), 8.42 (t, $J = 1.7$ Hz, 1H, Ar); ^{13}C-NMR (125 MHz, CDCl$_3$) δ: 21.0, 43.8, 45.0, 121.8, 121.9, 122.9, 124.9, 126.7, 129.8, 129.9, 130.0, 132.8, 140.7, 140.8, 146.0, 148.7, 152.8; *Anal.* calcd for C$_{17}$H$_{14}$N$_4$O$_2$S: C, 60.34; H, 4.17; N, 16.56. Found: C, 60.04; H, 4.13; N, 16.28.

3.1.49   Synthesis of (±)-3,4-Dihydro-9-[3-(1-hydroxyethyl)phenyl]-2H,6H-pyrimido [1,2-c][1,3]benzothiazin-6-imine (31l)

N-(*tert*-Butyl)-3,4-dihydro-9-(3-vinylphenyl)-2*H*,6*H*-pyrimido[1,2-c][1,3]benzothiazin-6-imine (29l). Using the general procedure as described for **22l**, *N*-(*tert*-butyl)-9-bromo-3,4-dihydro-2*H*,6*H*-pyrimido[1,2-c][1,3]benzothiazin-6-imine   **22k** (70.6 mg, 0.20 mmol) was allowed to react with 3-vinylphenylboronic acid (35.5 mg, 0.24 mmol) for 1 h. Purification by flash chromatography over aluminum oxide with *n*-hexane–EtOAc (1:0 to 9:1) gave the title compound **29l** as a colorless oil (64.4 mg, 86 %): IR (neat) cm^{-1}: 1591 (C=N); ^{1}H-NMR (500 MHz, CDCl$_3$) δ: 1.40 (s, 9H, 3 × CH$_3$), 1.90–1.95 (m, 2H, CH$_2$), 3.64 (t, $J = 5.4$ Hz, 2H, CH$_2$), 3.89 (t, $J = 6.3$ Hz, 2H, CH$_2$), 5.30 (d, $J = 10.9$ Hz, 1H, CH), 5.82 (d, $J = 17.8$ Hz, 1H, CH), 6.77 (dd, $J = 17.8$, 10.9 Hz, 1H, CH), 7.34 (d, $J = 1.7$ Hz, 1H, Ar), 7.37–7.47 (m, 4H, Ar), 7.59 (s, 1H, Ar), 8.25 (d, $J = 8.6$ Hz, 1H, Ar); ^{13}C-NMR (125 MHz, CDCl$_3$) δ: 21.9, 30.0 (3C), 45.1, 45.4, 54.2, 114.5, 122.7, 124.8, 125.0, 125.8, 126.5, 126.6, 128.9, 129.0, 129.5, 136.5, 138.2, 138.2, 139.7, 142.8, 147.7; HRMS (FAB): *m/z* calcd for C$_{23}$H$_{26}$N$_3$S [M + H]$^+$ 376.1847; found: 376.1850.

    Compound 31l. Using the general procedure as described for **25a**, compound **29l** (58.5 mg, 0.16 mmol) was allowed to react for 3 h. Purification by flash chromatography over aluminum oxide with *n*-hexane–EtOAc (9:1 to 1:1) gave the title compound **31l** as colorless solid (25.9 mg, 49 %): mp 193–195 °C (from CHCl$_3$–*n*-hexane); IR (neat) cm^{-1}: 1616 (C=N), 1558 (C=N); ^{1}H-NMR (400 MHz, DMSO-d_6) δ: 1.36 (d, $J = 6.6$ Hz, 3H, CH$_3$), 1.84–1.89 (m, 2H, CH$_2$), 3.59 (t, $J = 5.1$ Hz, 2H, CH$_2$), 3.92 (t, $J = 6.0$ Hz, 2H, CH$_2$), 4.76–4.81 (m, 1H, CH), 5.20 (d, $J = 4.4$ Hz, 1H, OH), 7.36–7.43 (m, 2H, Ar), 7.53–7.57 (m, 3H, Ar), 7.67 (s, 1H, Ar), 8.24 (d, $J = 8.3$ Hz, 1H, Ar), 8.71 (s, 1H, NH); ^{13}C-NMR (125 MHz, CDCl$_3$–CD$_3$OD) δ: 20.6, 25.0, 44.0, 44.5, 69.5, 121.7, 123.8, 125.0, 125.1, 125.3, 125.6, 128.8, 129.0, 129.0, 138.8, 143.6, 146.8, 147.2, 154.1; HRMS (FAB): *m/z* calcd for C$_{19}$H$_{20}$N$_3$OS [M + H]$^+$ 338.1327; found: 338.1327.

3.1.50 Synthesis of 9-[3-(Acetylamino)phenyl]-3,4-dihydro-2H,6H-pyrimido[1,2-c][1,3]benzothiazin-6-imine (31m)

9-[3-(Acetylamino)phenyl]-N-(tert-butyl)-3,4-dihydro-2H,6H-pyrimido[1,2-c] [1,3]benzothiazin-6-imine (29m). Using the general procedure as described for **22l**, N-(tert-butyl)-9-bromo-3,4-dihydro-2H,6H-pyrimido[1,2-c][1,3]benzothiazin-6-imine **22k** (52.8 mg, 0.15 mmol) was allowed to react with 3-(acetyl-amino)phenylboronic acid (32.2 mg, 0.18 mmol) in 1,4-dioxane for 1 h. Purification by flash chromatography over aluminum oxide with n-hexane–EtOAc (1:1) gave the title compound **29m** as colorless solid (44.6 mg, 73 %): mp 221–222 °C (from CHCl$_3$–n-hexane); IR (neat) cm^{-1}: 1670 (C=O), 1590 (C=N); ^{1}H-NMR (500 MHz, CDCl$_3$) δ: 1.40 (s, 9H, 3 × CH$_3$), 1.91–1.95 (m, 2H, CH$_2$), 2.18 (s, 3H, CH$_3$), 3.64 (t, J = 5.4 Hz, 2H, CH$_2$), 3.89 (t, J = 6.0 Hz, 2H, CH$_2$), 7.29–7.38 (m, 4H, Ar), 7.50 (d, J = 7.4 Hz, 1H, Ar), 7.70 (s, 1H, NH), 7.73 (s, 1H, Ar), 8.21 (d, J = 8.0 Hz, 1H, Ar); ^{13}C-NMR (125 MHz, CDCl$_3$) δ: 21.9, 24.5, 30.0 (3C), 45.1, 45.4, 54.2, 118.4, 119.5, 122.8, 122.9, 124.8, 126.7, 128.8, 129.5, 129.5, 138.2, 138.5, 140.2, 142.4, 147.9, 168.5; HRMS (FAB): m/z calcd for C$_{23}$H$_{27}$N$_4$OS [M + H]$^+$ 407.1906; found: 407.1905.

Compound 31m. Using the general procedure as described for **25a**, compound **29m** (35.4 mg, 0.096 mmol) was allowed to react for 3 h. Purification by pre-parative TLC over aluminum oxide with EtOAc–MeOH (98:2) gave the title compound **31m** as colorless solid (23.7 mg, 78 %): mp 208–210 °C (from MeOH–CHCl$_3$–n-hexane); IR (neat) cm^{-1}: 1691 (C=O), 1611 (C=N), 1561 (C=N); ^{1}H-NMR (DMSO-d$_6$) δ: 1.85-1.90 (m, 2H, CH$_2$), 2.06 (s, 3H, CH$_3$), 3.60 (t, J = 5.4 Hz, 2H, CH$_2$), 3.93 (t, J = 6.0 Hz, 2H, CH$_2$), 7.37–7.41 (m, 2H, Ar), 7.46–7.49 (m, 2H, Ar), 7.61 (d, J = 6.3 Hz, 1H, Ar), 7.90 (s, 1H, Ar), 8.26 (d, J = 8.0 Hz, 1H, Ar), 8.75 (s, 1H, NH), 10.04 (s, 1H, NH); ^{13}C-NMR (125 MHz, CDCl$_3$–CD$_3$OD) δ: 20.7, 23.7, 44.0, 44.5, 118.3, 119.6, 121.8, 122.4, 125.1 (2C), 129.0, 129.0, 129.2, 138.8, 139.4, 143.3, 147.2, 154.1, 169.7; HRMS (FAB): m/z calcd for C$_{19}$H$_{17}$N$_4$OS [M − H]$^-$ 349.1123; found: 349.1129.

3.1.51 Synthesis of 3,4-Dihydro-9-[3-(methanesulfonylamino)phenyl]-2H,6H-pyrimido[1,2-c][1,3]benzothiazin-6-imine (31n)

N-(tert-Butyl)-3,4-dihydro-9-[3-(methanesulfonylamino)phenyl]-2H,6H-py-rimido[1,2-c][1,3]benzothiazin-6-imine (29n). Using the general procedure as described for **22l**, N-(tert-butyl)-9-bromo-3,4-dihydro-2H,6H-pyrimido[1,2-c][1,3] benzothiazin-6-imine **22k** (52.8 mg, 0.15 mmol) was allowed to react with 3-(methanesulfonylamino)phenylboronic acid (38.7 mg, 0.18 mmol) in 1,4-dioxane for 1 h. Purification by flash chromatography over aluminum oxide with n-hexane–EtOAc (1:2 to 0:1) gave the title compound **29n** as colorless solid (20.1 mg, 30 %): mp 200–202 °C (from CHCl$_3$–n-hexane); IR (neat) cm^{-1}: 1591 (C=N), 1153 (NSO$_2$); ^{1}H-NMR (500 MHz, CDCl$_3$–CD$_3$OD) δ: 1.41 (s, 9H, 3 × CH$_3$), 1.92-1.97 (m, 2H, CH$_2$), 3.02 (s, 3H, CH$_3$), 3.62 (t, J = 5.2 Hz, 2H, CH$_2$), 3.90 (t, J = 6.0 Hz, 2H, CH$_2$), 7.27 (d, J = 7.4 Hz, 1H, Ar), 7.34-7.43 (m, 5H, Ar), 8.15 (d, J = 8.6 Hz, 1H, Ar); ^{13}C-NMR (125 MHz, CDCl$_3$) δ: 21.8, 30.0 (3C), 39.5, 44.9, 45.5, 54.3, 119.1, 119.8, 122.9, 123.7, 124.8, 127.0, 129.0, 129.8, 130.1, 137.7, 137.9, 140.9, 142.0, 148.4; HRMS (FAB): m/z calcd for C$_{22}$H$_{27}$N$_4$O$_2$S$_2$ [M + H]$^+$ 443.1575; found: 443.1574.

Compound 31n. Using the general procedure as described for **25a**, compound **29n** (26.1 mg, 0.059 mmol) was allowed to react for 4.5 h. Purification by preparative TLC over aluminum oxide with EtOAc–MeOH (95:5) gave the title compound **31n** as colorless solid (13.4 mg, 59 %): mp 194–196 °C (from MeOH–CHCl$_3$–n-hexane); IR (neat) cm^{-1}: 1625 (C=N), 1557 (C=N), 1325 (NSO$_2$), 1147 (NSO$_2$); ^{1}H-NMR (500 MHz, CDCl$_3$–CD$_3$OD) δ: 1.98-2.03 (m, 2H, CH$_2$), 3.02 (s, 3H, CH$_3$), 3.70 (t, J = 5.4 Hz, 2H, CH$_2$), 4.00 (t, J = 6.3 Hz, 2H, CH$_2$), 7.23–7.46 (m, 6H, Ar), 8.21 (d, J = 8.6 Hz, 1H, Ar); ^{13}C-NMR (125 MHz, CDCl$_3$–CD$_3$OD) δ: 20.8, 39.1, 44.0, 44.7, 118.8, 120.0, 121.9, 123.4, 125.2, 125.6, 129.3, 129.5, 130.1, 138.0, 140.5, 142.7, 147.0, 153.9; HRMS (FAB): m/z calcd for C$_{18}$H$_{19}$N$_4$O$_2$S$_2$ [M + H]$^+$ 387.0949; found: 387.0957.

3.1.52   Synthesis of 3,4-Dihydro-9-(3-hydroxyphenyl)-2H,6H-pyrimido[1,2-c][1,3]benzothiazin-6-imine (31o)

N-($tert$-Butyl)-3,4-dihydro-9-(3-hydroxyphenyl)-2H,6H-pyrimido[1,2-c][1,3]benzothiazin-6-imine (29o). Using the general procedure as described for **22l**, N-($tert$-butyl)-9-bromo-3,4-dihydro-2H,6H-pyrimido[1,2-c][1,3]benzothiazin-6-imine **22k** (52.8 mg, 0.15 mmol) was allowed to react with 3-hydroxyphenylboronic acid (24.8 mg, 0.18 mmol) in 1,4-dioxane for 1 h. Purification by flash chromatography over aluminum oxide with n-hexane–EtOAc (1:1 to 0:1) gave the title compound **29o** as colorless solid (16.1 mg, 29 %): mp 265–267 °C (from MeOH–CHCl$_3$–n-hexane); IR (neat) cm^{-1}: 1591 (C=N); ^{1}H-NMR (500 MHz, CDCl$_3$–CD$_3$OD) δ: 1.41 (s, 9H, 3 × CH$_3$), 1.92–1.97 (m, 2H, CH$_2$), 3.60 (t, J = 5.2 Hz, 2H, CH$_2$), 3.90 (t, J = 6.0 Hz, 2H, CH$_2$), 6.85 (d, J = 8.0 Hz, 1H, Ar), 7.03 (s, 1H, Ar), 7.08 (d, J = 8.0 Hz, 1H, Ar), 7.27 (t, J = 8.0 Hz, 1H, Ar), 7.34 (s, 1H, Ar), 7.43 (d, J = 8.0 Hz, 1H, Ar), 8.09 (d, J = 8.0 Hz, 1H, Ar); ^{13}C-NMR (125 MHz, CDCl$_3$–CD$_3$OD) δ: 21.5, 29.7 (3C), 44.5, 45.4, 54.2, 113.8, 115.1, 118.2, 122.8, 124.9, 126.0, 128.5, 129.4, 129.8, 138.1, 140.4, 143.2, 149.4, 157.2; HRMS (FAB): m/z calcd for C$_{21}$H$_{24}$N$_3$OS [M + H]$^+$ 366.1640; found: 366.1639.

Compound 31o. TFA (9 mL) was added to a mixture of **29o** (16.1 mg, 0.044 mmol) and MS4Å (2.0 g, powder, activated by heating with Bunsen burner) in CHCl$_3$ (1.0 mL) and MeOH (10 drops). After being stirred under reflux for 5 h, the mixture was concentrated. To a mixture of the residue in CHCl$_3$ was added dropwise Et$_3$N at 0 °C to adjust pH to 8–9. The whole was extracted with EtOAc. The extract was washed with sat. NaHCO$_3$, brine, and dried over MgSO$_4$. After concentration, the residue was purified by preparative TLC over aluminum oxide with EtOAc–MeOH (95:5) to give the title compound **31o** as colorless solid (7.9 mg, 58 %): mp 199–200 °C (from MeOH–CHCl$_3$–n-hexane); IR (neat) cm^{-1}: 1621 (C=N), 1557 (C=N); ^{1}H-NMR (500 MHz, DMSO-d_6) δ: 1.85-1.90 (m, 2H, CH$_2$), 3.60 (t, J = 5.4 Hz, 2H, CH$_2$), 3.92 (t, J = 5.4 Hz, 2H, CH$_2$), 6.81 (dd, J = 7.7, 2.0 Hz, 1H, Ar), 7.06 (t, J = 2.0 Hz, 1H, Ar), 7.13 (d, J = 8.3 Hz, 1H, Ar), 7.27 (t, J = 7.7 Hz, 1H, Ar), 7.48–7.50 (m, 2H, Ar), 8.22 (d, J = 8.3 Hz, 1H, Ar), 8.74 (br s, 1H, NH), 9.57 (s, 1H, OH); ^{13}C-NMR (100 MHz, DMSO-d_6) δ: 20.7, 43.1, 44.3, 113.5, 115.3, 117.5, 121.4, 124.2, 124.9, 129.0, 129.5, 130.0, 139.5, 142.3, 145.2, 149.7, 157.8; HRMS (FAB): m/z calcd for C$_{17}$H$_{16}$N$_3$OS [M + H]$^+$ 310.1014; found: 310.1010.

3.1.53 Synthesis of 3,4-Dihydro-9-(4-methoxyphenyl)-2H,6H-pyrimido[1,2-c][1,3]benzothiazin-6-imine (31p)

N-(*tert*-Butyl)-3,4-dihydro-9-(3-methoxyphenyl)-2H,6H-pyrimido[1,2-c][1,3]benzothiazin-6-imine (29p). Using the general procedure as described for **22l**, N-(*tert*-butyl)-9-bromo-3,4-dihydro-2H,6H-pyrimido[1,2-c][1,3]benzothiazin-6-imine **22k** (52.8 mg, 0.15 mmol) was allowed to react with 4-methoxyphenylboronic acid (27.4 mg, 0.18 mmol) for 1 h. Purification by flash chromatography over aluminum oxide with n-hexane–EtOAc (1:0 to 9:1) gave the title compound **29p** as a colorless oil (55.6 mg, 98 %): IR (neat) cm^{-1}: 1591 (C=N); ^{1}H-NMR (400 MHz, CDCl$_3$) δ: 1.40 (s, 9H, 3 × CH$_3$), 1.90–1.96 (m, 2H, CH$_2$), 3.64 (t, J = 5.5 Hz, 2H, CH$_2$), 3.86 (s, 3H, CH$_3$), 3.89 (t, J = 6.2 Hz, 2H, CH$_2$), 6.91 (dd, J = 8.2, 2.5 Hz, 1H, Ar), 7.10 (t, J = 2.5 Hz, 1H, Ar), 7.15–7.18 (m, 1H, Ar), 7.32–7.37 (m, 2H, Ar), 7.41 (dd, J = 8.5, 1.7 Hz, 1H, Ar), 8.24 (d, J = 8.5 Hz, 1H, Ar); ^{13}C-NMR (100 MHz, CDCl$_3$) δ: 21.9, 30.0 (3C), 45.1, 45.4, 54.2, 55.3, 112.7, 113.5, 119.5, 122.8, 124.9, 126.7, 128.9, 129.5, 129.9, 138.2, 140.9, 142.8, 147.7, 160.0; HRMS (FAB): m/z calcd for C$_{22}$H$_{26}$N$_3$OS [M + H]$^+$ 380.1797; found: 380.1793.

Compound 31p. Using the general procedure as described for **25a**, compound **29p** (32.1 mg, 0.085 mmol) was allowed to react for 2 h with TFA (1.0 mL) and MS4Å (150 mg). Purification by flash chromatography over aluminum oxide with n-hexane–EtOAc (3:1) gave the title compound **31p** as colorless solid (23.4 mg, 85 %): mp 114–115 °C (from CHCl$_3$–n-hexane); IR (neat) cm^{-1}: 1620 (C=N), 1569 (C=N); ^{1}H-NMR (400 MHz, CDCl$_3$) δ: 1.96–2.02 (m, 2H, CH$_2$), 3.71 (t,

$J = 5.5$ Hz, 2H, CH$_2$), 3.86 (s, 3H, CH$_3$), 4.04 (t, $J = 6.2$ Hz, 2H, CH$_2$), 6.93 (dd, $J = 7.9$, 2.7 Hz, 1H, Ar), 7.09 (t, $J = 2.7$ Hz, 1H, Ar), 7.14–7.17 (m, 1H, Ar), 7.21 (br s, NH), 7.24 (d, $J = 1.7$ Hz, 1H, Ar), 7.36 (t, $J = 7.9$ Hz, 1H, Ar), 7.44 (dd, $J = 8.5$, 1.7 Hz, 1H, Ar), 8.27 (d, $J = 8.5$ Hz, 1H, Ar); ^{13}C-NMR (100 MHz, CDCl$_3$) δ: 21.0, 43.8, 45.0, 55.4, 112.8, 113.6, 119.5, 121.8, 125.2, 125.6, 129.3, 129.3, 129.9, 140.6, 143.4, 146.4, 153.4, 160.0; *Anal.* calcd for C$_{18}$H$_{17}$N$_3$OS: C, 66.85; H, 5.30; N, 12.99. Found: C, 66.56; H, 5.14; N, 12.83.

3.1.54 Synthesis of 3,4-Dihydro-9-(3-isopropoxyphenyl)-2H,6H-pyrimido[1,2-c][1,3]benzothiazin-6-imine (31q)

N-(tert-Butyl)-3,4-dihydro-9-(3-isopropoxyphenyl)-2H,6H-pyrimido[1,2-c] [1, 3]benzothiazin-6-imine (29q). Using the general procedure as described for **22l**, N-(tert-butyl)-9-bromo-3,4-dihydro-2H,6H-pyrimido[1,2-c][1,3]benzothiazin-6-imine **22k** (52.8 mg, 0.15 mmol) was allowed to react with 3-isopropoxyphenylboronic acid (32.4 mg, 0.18 mmol) for 1 h. Purification by flash chromatography over aluminum oxide with n-hexane–EtOAc (1:0 to 9:1) gave the title compound **29q** as a colorless oil (48.8 mg, 80 %): IR (neat) cm^{-1}: 1592 (C=N); ^{1}H-NMR (500 MHz, CDCl$_3$) δ: 1.36 (d, $J = 5.7$ Hz, 6H, 2 × CH$_3$), 1.40 (s, 9H, 3 × CH$_3$), 1.90-1.95 (m, 2H, CH$_2$), 3.64 (t, $J = 5.7$ Hz, 2H, CH$_2$), 3.89 (t, $J = 6.0$ Hz, 2H, CH$_2$), 4.58-4.65 (m, 1H, CH), 6.89 (dd, $J = 8.0$, 2.0 Hz, 1H, Ar), 7.10 (t, $J = 2.0$ Hz, 1H, Ar), 7.14 (dd, $J = 7.4$, 2.0 Hz, 1H, Ar), 7.31–7.34 (m, 2H, Ar), 7.41 (dd, $J = 8.0$, 1.7 Hz, 1H, Ar), 8.24 (d, $J = 8.0$ Hz, 1H, Ar); ^{13}C-NMR (125 MHz, CDCl$_3$) δ: 21.9, 22.0 (2C), 30.0 (3C), 45.1, 45.4, 54.1, 69.9, 114.9, 115.1, 119.3, 122.8, 124.8, 126.6, 128.8, 129.4, 129.9, 138.3, 140.9, 142.8, 147.7, 158.3; HRMS (FAB): m/z calcd for C$_{24}$H$_{30}$N$_3$OS [M + H]$^+$ 408.2110; found: 408.2108.

Compound 31q. Using the general procedure as described for **25a**, compound **29q** (44.0 mg, 0.108 mmol) was allowed to react for 2 h. Purification by flash chromatography over aluminum oxide with n-hexane–EtOAc (9:1 to 7:3) gave the title compound **31q** as a colorless oil (35.5 mg, 94 %): IR (neat) cm^{-1}: 1620 (C=N), 1567 (C=N); ^{1}H-NMR (400 MHz, CDCl$_3$) δ: 1.36 (d, $J = 6.1$ Hz, 6H, 2 × CH$_3$), 1.95–2.01 (m, 2H, CH$_2$), 3.70 (t, $J = 5.5$ Hz, 2H, CH$_2$), 4.03 (t, $J = 6.2$ Hz, 2H, CH$_2$), 4.57–4.66 (m, 1H, CH), 6.90 (dd, $J = 8.3$, 2.4 Hz, 1H, Ar), 7.09 (d, $J = 1.8$ Hz, 1H, Ar), 7.11–7.13 (m, 1H, Ar), 7.23 (d, $J = 1.8$ Hz, 1H, Ar), 7.33 (t, $J = 8.0$ Hz, 1H, Ar), 7.43 (dd, $J = 8.5$, 1.8 Hz, 1H, Ar), 8.26 (d, $J = 8.5$ Hz, 1H, Ar); ^{13}C-NMR (100 MHz, CDCl$_3$) δ: 21.0, 22.0 (2C), 43.8, 45.0, 70.0, 115.0, 115.2, 119.2, 121.8, 125.1, 125.5, 129.2, 129.3, 129.9, 140.6, 143.4, 146.4, 153.4, 158.3; HRMS (FAB): m/z calcd for C$_{20}$H$_{22}$N$_3$OS [M + H]$^+$ 352.1484; found: 352.1484.

3.1.55 Synthesis of 9-[(1,1'-Biphenyl)-3-yl]-3,4-dihydro-2H,6H-pyrimido[1,2-c][1,3]benzothiazin-6-imine (31r)

9-[(1,1'-Biphenyl)-3-yl]-N-(tert-butyl)-3,4-dihydro-2H,6H-pyrimido[1,2-c][1,3] benzothiazin-6-imine (29r). Using the general procedure as described for **22l**, N-(tert-butyl)-9-bromo-3,4-dihydro-2H,6H-pyrimido[1,2-c][1,3]benzothiazin-6-imine **22k** (52.8 mg, 0.15 mmol) was allowed to react with 3-biphenylboronic acid (35.7 mg, 0.18 mmol) for 1 h. Purification by flash chromatography over aluminum oxide with n-hexane–EtOAc (1:0 to 9:1) gave the title compound **29r** as a colorless oil (65.1 mg, > 99 %.): IR (neat) cm^{-1}: 1591 (C=N); ^{1}H-NMR (500 MHz, CDCl$_3$) δ: 1.41 (s, 9H, 3 × CH$_3$), 1.91–1.95 (m, 2H, CH$_2$), 3.64 (t, J = 5.4 Hz, 2H, CH$_2$), 3.89 (t, J = 6.3 Hz, 2H, CH$_2$), 7.35–7.39 (m, 2H, Ar), 7.44–7.51 (m, 4H, Ar), 7.55–7.60 (m, 2H, Ar), 7.63 (d, J = 7.4 Hz, 2H, Ar), 7.78 (s, 1H, Ar), 8.27 (d, J = 8.0 Hz, 1H, Ar); ^{13}C-NMR (125 MHz, CDCl$_3$) δ: 21.9, 30.0 (3C), 45.1, 45.4, 54.2, 122.8, 124.9, 125.9 (2C), 126.7, 126.9, 127.2 (2C), 127.5, 128.8 (2C), 129.0, 129.3, 129.6, 138.2, 139.9, 140.9, 142.0, 142.8, 147.7; HRMS (FAB): m/z calcd for C$_{27}$H$_{28}$N$_3$S [M + H]$^+$ 426.2004; found: 426.2000.

Compound 31r. Using the general procedure as described for **25a**, compound **29r** (56.1 mg, 0.13 mmol) was allowed to react for 3 h. Purification by flash chromatography over aluminum oxide with n-hexane–EtOAc (9:1 to 7:3) gave the title compound **31r** as colorless solid (37.0 mg, 76 %): mp 179–181 °C (from CHCl$_3$–n-hexane); IR (neat) cm^{-1}: 1620 (C=N), 1568 (C=N); ^{1}H-NMR (500 MHz, CDCl$_3$) δ: 1.97–2.02 (m, 2H, CH$_2$), 3.72 (t, J = 5.4 Hz, 2H, CH$_2$), 4.04 (t, J = 6.0 Hz, 2H, CH$_2$), 7.30 (d, J = 1.7 Hz, 1H, Ar), 7.38 (t, J = 7.4 Hz, 1H, Ar), 7.45–7.56 (m, 5H, Ar), 7.59–7.64 (m, 3H, Ar), 7.77 (s, 1H, Ar), 8.30 (d, J = 8.6 Hz, 1H, Ar); ^{13}C-NMR (125 MHz, CDCl$_3$) δ: 21.0, 43.9, 45.0, 121.9, 125.2, 125.6, 125.9 (2C), 127.1, 127.2 (2C), 127.6, 128.8 (2C), 129.3, 129.4 (2C), 139.7, 140.8, 142.1, 143.4, 146.4, 153.3; HRMS (FAB): m/z calcd for C$_{23}$H$_{20}$N$_3$S [M + H]$^+$ 370.1378; found: 370.1378.

3.1.56 Synthesis of 3,4-Dihydro-9-(2-methoxyphenyl)-2H,6H-pyrimido[1,2-c][1,3]benzothiazin-6-imine (31s)

N-(tert-Butyl)-3,4-dihydro-9-(2-methoxyphenyl)-2H,6H-pyrimido[1,2-c][1,3] benzothiazin-6-imine (29s). Using the general procedure as described for **22l**, N-(tert-butyl)-9-bromo-3,4-dihydro-2H,6H-pyrimido[1,2-c][1,3]benzothiazin-6-imine **22k** (52.8 mg, 0.15 mmol) was allowed to react with 2-methoxyphenylboronic acid (27.4 mg, 0.18 mmol) for 1 h. Purification by flash chromatography over aluminum oxide with n-hexane–EtOAc (1:0 to 9:1) gave the title compound **29s** as a colorless oil (46.0 mg, 81 %): IR (neat) cm^{-1}: 1591 (C=N); ^{1}H-NMR (400 MHz, CDCl$_3$) δ: 1.39 (s, 9H, 3 × CH$_3$), 1.89–1.95 (m, 2H, CH$_2$), 3.63 (t, J = 5.4 Hz, 2H, CH$_2$), 3.80 (s, 3H, CH$_3$), 3.88 (t, J = 5.9 Hz, 2H, CH$_2$), 6.97 (d, J = 8.3 Hz, 1H, Ar), 7.02 (t, J = 7.4 Hz, 1H, Ar), 7.29–7.38 (m, 4H, Ar), 8.21 (d, J = 8.3 Hz, 1H, Ar); ^{13}C-NMR (100 MHz, CDCl$_3$) δ: 22.0, 30.0 (3C), 45.1, 45.4, 54.1, 55.6, 111.4, 120.9, 125.2, 126.2, 127.5, 128.0, 128.6, 129.1, 129.3, 130.6, 138.6, 140.6, 147.9, 156.5; HRMS (FAB): m/z calcd for C$_{22}$H$_{26}$N$_3$OS [M + H]$^+$ 380.1797; found: 380.1793.

Compound 31s. Using the general procedure as described for **25a**, compound **29s** (34.7 mg, 0.091 mmol) was allowed to react for 2 h. Purification by flash chromatography over aluminum oxide with n-hexane–EtOAc (9:1 to 7:3) gave the title compound **31s** as colorless solid (21.6 mg, 73 %): mp 130.5 °C (from CHCl$_3$–n-hexane); IR (neat) cm^{-1}: 1620 (C=N), 1567 (C=N); ^{1}H-NMR (400 MHz, CDCl$_3$) δ: 1.95-2.01 (m, 2H, CH$_2$), 3.70 (t, J = 5.6 Hz, 2H, CH$_2$), 3.80 (s, 3H, CH$_3$), 4.03 (t, J = 6.2 Hz, 2H, CH$_2$), 6.97 (d, J = 8.3 Hz, 1H, Ar), 7.02 (td, J = 7.5, 0.8 Hz, 1H, Ar), 7.22 (d, J = 1.5 Hz, 1H, Ar), 7.28-7.40 (m, 3H, Ar), 8.24 (d, J = 8.3 Hz, 1H, Ar); ^{13}C-NMR (100 MHz, CDCl$_3$) δ: 21.1, 43.8, 44.9, 55.6, 111.3, 120.9, 124.3, 125.1, 127.7, 128.3, 128.4, 128.7, 129.5, 130.5, 141.1, 146.6, 153.7, 156.4; *Anal.* calcd for C$_{18}$H$_{17}$N$_3$OS: C, 66.85; H, 5.30; N, 12.99. Found: C, 66.56; H, 5.08; N, 12.90.

3.1.57 Synthesis of 9-[(1,1′-Biphenyl)-2-yl]-3,4-dihydro-2H,6H-pyrimido[1,2-c][1,3]benzothiazin-6-imine (31t)

9-[(1,1′-Biphenyl)-2-yl]-N-(*tert*-butyl)-3,4-dihydro-2H,6H-pyrimido[1,2-c][1,3] benzothiazin-6-imine (29t). Using the general procedure as described for **22l**, N-(*tert*-butyl)-9-bromo-3,4-dihydro-2H,6H-pyrimido[1,2-c][1,3]benzothiazin-6-imine **22k** (52.8 mg, 0.15 mmol) was allowed to react with 2-biphenylboronic acid (35.7 mg, 0.18 mmol) for 1 h. Purification by flash chromatography over aluminum oxide with n-hexane–EtOAc (1:0 to 9:1) gave the title compound **29t** as a colorless oil (64.1 mg, >99 %.): IR (neat) cm^{-1}: 1591 (C=N); ^{1}H-NMR (400 MHz, CDCl$_3$) δ: 1.36 (s, 9H, 3 × CH$_3$), 1.86–1.91 (m, 2H, CH$_2$), 3.59 (t, J = 5.6 Hz, 2H, CH$_2$), 3.85 (t, J = 6.2 Hz, 2H, CH$_2$), 6.91–6.93 (m, 2H, Ar), 7.11–7.14 (m, 2H, Ar), 7.19–7.22 (m, 3H, Ar), 7.40–7.42 (m, 4H, Ar), 7.99 (d, J = 8.8 Hz, 1H, Ar); ^{13}C-NMR (100 MHz, CDCl$_3$) δ: 21.9, 29.9 (3C), 45.1, 45.4, 54.1, 125.5, 125.9, 126.7, 127.5, 127.9, 127.9, 128.1, 128.1 (2C), 128.6, 129.7 (2C), 130.2, 130.7, 138.4, 138.9, 140.6, 140.9, 143.6, 147.8; HRMS (FAB): m/z calcd for C$_{27}$H$_{28}$N$_3$S [M + H]$^+$ 426.2004; found: 426.2002.

Compound 31t. Using the general procedure as described for **25a**, compound **29t** (51.1 mg, 0.12 mmol) was allowed to react for 3 h. Purification by flash chromatography over aluminum oxide with n-hexane–EtOAc (9:1 to 7:3) gave the title compound **31t** as a colorless oil (35.7 mg, 81 %): IR (neat) cm^{-1}: 1620

(C=N), 1570 (C=N); ^{1}H-NMR (400 MHz, CDCl$_3$) δ: 1.92-1.98 (m, 2H, CH$_2$), 3.66 (t, $J = 5.5$ Hz, 2H, CH$_2$), 3.99 (t, $J = 6.1$ Hz, 2H, CH$_2$), 6.83 (d, $J = 1.7$ Hz, 1H, Ar), 6.95 (dd, $J = 8.3$, 1.7 Hz, 1H, Ar), 7.11–7.13 (m, 2H, Ar), 7.20–7.23 (m, 3H, Ar), 7.38–7.43 (m, 4H, Ar), 8.03 (d, $J = 8.3$ Hz, 1H, Ar); ^{13}C-NMR (100 MHz, CDCl$_3$) δ: 21.0, 43.8, 44.9, 124.5, 124.8, 126.8, 127.6, 128.1 (3C), 128.1, 128.2, 128.3, 128.4, 129.7 (2C), 130.2, 138.5, 140.6, 140.8, 144.2, 146.5, 153.5; HRMS (FAB): m/z calcd for C$_{23}$H$_{20}$N$_3$S [M + H]$^+$ 370.1378; found: 370.1378.

3.1.58 Synthesis of 3,4-Dihydro-9-(3,4-dimethoxyphenyl)-2H,6H-pyrimido[1,2-c][1,3]benzothiazin-6-imine (31u)

N-(*tert*-Butyl)-3,4-dihydro-9-(3,4-dimethoxyphenyl)-2H,6H-pyrimido[1,2-c][1,3]benzothiazin-6-imine (29u). Using the general procedure as described for **22l**, N-(*tert*-butyl)-9-bromo-3,4-dihydro-2H,6H-pyrimido[1,2-c][1,3]benzothiazin-6-imine **22k** (52.8 mg, 0.15 mmol) was allowed to react with 3,4-dimethoxyphenylboronic acid (32.8 mg, 0.18 mmol) for 1 h. Purification by flash chromatography over aluminum oxide with n-hexane–EtOAc (1:0 to 9:1) gave the title compound **29u** as colorless solid (60.3 mg, 98 %): mp 147–148 °C (from CHCl$_3$–n-hexane); IR (neat) cm^{-1}: 1593 (C=N); ^{1}H-NMR (400 MHz, CDCl$_3$) δ: 1.40 (s, 9H, 3 × CH$_3$), 1.90–1.96 (m, 2H, CH$_2$), 3.64 (t, $J = 5.5$ Hz, 2H, CH$_2$), 3.89 (t, $J = 6.1$ Hz, 2H, CH$_2$), 3.92 (s, 3H, CH$_3$), 3.96 (s, 3H, CH$_3$), 6.93 (d, $J = 8.3$ Hz, 1H, Ar), 7.09 (d, $J = 2.0$ Hz, 1H, Ar), 7.15 (dd, $J = 8.3$, 2.0 Hz, 1H, Ar), 7.29 (d, $J = 1.7$ Hz, 1H, Ar), 7.39 (dd, $J = 8.3$, 1.7 Hz, 1H, Ar), 8.23 (d, $J = 8.3$ Hz, 1H, Ar); ^{13}C-NMR (125 MHz, CDCl$_3$) δ: 21.9, 30.0 (3C), 45.1, 45.4, 54.1, 56.0, 56.0, 110.1, 111.4, 119.5, 122.2, 124.5, 126.1, 128.9, 129.4, 132.3, 138.3, 142.7, 147.7, 149.3 (2C); HRMS (FAB): m/z calcd for C$_{23}$H$_{28}$N$_3$O$_2$S [M + H]$^+$ 410.1902; found: 410.1907.

Compound 31u. Using the general procedure as described for **25a**, compound **29u** (46.0 mg, 0.11 mmol) was allowed to react for 3 h. Purification by flash chromatography over aluminum oxide with n-hexane–EtOAc (9:1 to 7:3) gave the title compound **31u** as colorless solid (24.6 mg, 63 %): mp 142 °C (from CHCl$_3$–n-hexane); IR (neat) cm^{-1}: 1620 (C=N), 1567 (C=N); ^{1}H-NMR (400 MHz, CDCl$_3$) δ: 1.96-2.02 (m, 2H, CH$_2$), 3.71 (t, $J = 5.6$ Hz, 2H, CH$_2$), 3.92 (s, 3H, CH$_3$), 3.96 (s, 3H, CH$_3$), 4.04 (t, $J = 6.2$ Hz, 2H, CH$_2$), 6.93 (d, $J = 8.3$ Hz, 1H, Ar), 7.07 (d, $J = 2.0$ Hz, 1H, Ar), 7.14 (dd, $J = 8.3$, 2.0 Hz, 1H, Ar), 7.20 (d, $J = 1.8$ Hz, 1H, Ar), 7.41 (dd, $J = 8.3$, 1.8 Hz, 1H, Ar), 8.25 (d, $J = 8.3$ Hz, 1H, Ar); ^{13}C-NMR (100 MHz, CDCl$_3$) δ: 21.0, 43.8, 44.9, 56.0, 56.0, 110.1, 111.5, 119.5, 121.3, 124.7, 125.0, 129.3, 129.4, 132.0, 143.2, 146.4, 149.3, 149.4, 153.4; Anal. calcd for C$_{19}$H$_{19}$N$_3$O$_2$S: C, 64.57; H, 5.42; N, 11.89. Found: C, 64.41; H, 5.37; N, 11.93.

3.1.59 Synthesis of 3,4-Dihydro-9-(3,4,5-trimethoxyphenyl)-2H,6H-pyrimido[1,2-c][1,3]benzothiazin-6-imine (31v)

N-(tert-Butyl)-3,4-dihydro-9-(3,4,5-trimethoxyphenyl)-2H,6H-pyrimido[1,2-c][1,3]benzothiazin-6-imine (29v). Using the general procedure as described for **22l**, N-(tert-butyl)-9-bromo-3,4-dihydro-2H,6H-pyrimido[1,2-c][1,3]benzothiazin-6-imine **22k** (52.8 mg, 0.15 mmol) was allowed to react with 3,4,5-trimethoxyphenylboronic acid (38.2 mg, 0.18 mmol) for 1 h. Purification by flash chromatography over aluminum oxide with n-hexane–EtOAc (1:0 to 9:1) gave the title compound **29v** as a colorless oil (65.0 mg, 99 %): IR (neat) cm^{-1}: 1585 (C=N); ^{1}H-NMR (400 MHz, CDCl$_3$) δ: 1.41 (s, 9H, 3 × CH$_3$), 1.91–1.96 (m, 2H, CH$_2$), 3.64 (t, J = 5.5 Hz, 2H, CH$_2$), 3.88–3.91 (m, 5H, CH$_3$, CH$_2$), 3.93 (s, 6H, 2 × CH$_3$), 6.77 (s, 2H, Ar), 7.28 (d, J = 1.8 Hz, 1H, Ar), 7.38 (dd, J = 8.4, 1.8 Hz, 1H, Ar), 8.24 (d, J = 8.4 Hz, 1H, Ar); ^{13}C-NMR (100 MHz, CDCl$_3$) δ: 21.9, 30.0 (3C), 45.1, 45.4, 54.2, 56.3 (2C), 60.9, 104.4 (2C), 122.6, 124.7, 126.5, 128.9, 129.5, 135.3, 138.1, 138.3, 143.0, 147.7, 153.6 (2C); HRMS (FAB): m/z calcd for C$_{24}$H$_{30}$N$_3$O$_3$S [M + H]$^+$ 440.2008; found: 440.2008.

Compound 31v. Using the general procedure as described for **25a**, compound **29v** (49.4 mg, 0.11 mmol) was allowed to react for 3 h. Purification by flash chromatography over aluminum oxide with n-hexane–EtOAc (9:1 to 7:3) gave the title compound **31v** as colorless solid (14.2 mg, 34 %): mp 156–157 °C (from CHCl$_3$–n-hexane); IR (neat) cm^{-1}: 1620 (C=N), 1569 (C=N); ^{1}H-NMR (400 MHz, CDCl$_3$) δ: 1.97–2.02 (m, 2H, CH$_2$), 3.71 (t, J = 5.5 Hz, 2H, CH$_2$), 3.89 (s, 3H, CH$_3$), 3.93 (s, 6H, 2 × CH$_3$), 4.04 (t, J = 6.1 Hz, 2H, CH$_2$), 6.75 (s, 2H, Ar), 7.20 (d, J = 1.7 Hz, 1H, Ar), 7.40 (dd, J = 8.5, 1.7 Hz, 1H, Ar), 8.27 (d, J = 8.5 Hz, 1H, Ar); ^{13}C-NMR (100 MHz, CDCl$_3$) δ: 21.0, 43.9, 45.0, 56.3 (2C), 61.0, 104.4 (2C), 121.7, 125.0, 125.4, 129.3 (2C), 135.0, 138.5, 143.6, 146.4, 153.3, 153.6 (2C); HRMS (FAB): m/z calcd for C$_{20}$H$_{22}$N$_3$O$_3$S [M + H]$^+$ 384.1382; found: 384.1381.

3.1.60 Synthesis of 9-(3-Chloro-4-methoxyphenyl)-3,4-dihydro-2H,6H-pyrimido[1,2-c][1,3]benzothiazin-6-imine (31w)

N-(tert-Butyl)-9-(3-chloro-4-methoxyphenyl)-3,4-dihydro-2H,6H-pyrimido[1,2-c][1,3]benzothiazin-6-imine (29w). Using the general procedure as described for **22l**, N-(tert-butyl)-9-bromo-3,4-dihydro-2H,6H-pyrimido[1,2-c] [1,3]benzothiazin-6-imine **22k** (52.8 mg, 0.15 mmol) was allowed to react with 3-chloro-4-methoxyphenylboronic acid (33.6 mg, 0.18 mmol) for 1 h. Purification by flash chromatography over aluminum oxide with n-hexane–EtOAc (1:0 to 9:1) gave the title compound **29w** as a colorless oil (58.7 mg, 95 %): IR (neat) cm^{-1}: 1592 (C=N); ^{1}H-NMR (500 MHz, CDCl$_3$) δ: 1.40 (s, 9H, 3 × CH$_3$), 1.90–1.95 (m, 2H, CH$_2$), 3.64 (t, J = 5.4 Hz, 2H, CH$_2$), 3.89 (t, J = 6.0 Hz, 2H, CH$_2$), 3.94 (s, 3H, CH$_3$), 6.98 (d, J = 8.6 Hz, 1H, Ar), 7.26 (d, J = 2.0 Hz, 1H, Ar), 7.36 (dd, J = 8.6, 2.0 Hz, 1H, Ar), 7.45 (dd, J = 8.6, 2.3 Hz, 1H, Ar), 7.61 (d, J = 2.3 Hz, 1H, Ar), 8.23 (d, J = 8.6 Hz, 1H, Ar); ^{13}C-NMR (125 MHz, CDCl$_3$) δ: 21.9, 30.0 (3C), 45.1, 45.4, 54.2, 56.2, 112.2, 122.2, 123.0, 124.3, 126.2, 126.4, 128.7, 129.0, 129.7, 132.7, 138.1, 141.2, 147.6, 155.0; HRMS (FAB): m/z calcd for C$_{22}$H$_{25}$ClN$_3$OS [M + H]$^+$ 414.1407; found: 414.1402.

Compound 31w. Using the general procedure as described for **25a**, compound **29w** (41.6 mg, 0.10 mmol) was allowed to react for 3 h. Purification by flash chromatography over aluminum oxide with *n*-hexane–EtOAc (9:1 to 7:3) gave the title compound **31w** as colorless solid (18.8 mg, 53 %): mp 186–188 °C (from CHCl$_3$–*n*-hexane); IR (neat) cm^{-1}: 1620 (C=N), 1563 (C=N); ^{1}H-NMR (500 MHz, CDCl$_3$) δ: 1.97–2.01 (m, 2H, CH$_2$), 3.71 (t, J = 5.4 Hz, 2H, CH$_2$), 3.94 (s, 3H, CH$_3$), 4.03 (t, J = 6.0 Hz, 2H, CH$_2$), 6.98 (d, J = 8.6 Hz, 1H, Ar), 7.17 (d, J = 1.4 Hz, 1H, Ar), 7.38 (dd, J = 8.0, 2.3 Hz, 1H, Ar), 7.44 (dd, J = 8.0, 1.4 Hz, 1H, Ar), 7.59 (d, J = 2.3 Hz, 1H, Ar), 8.25 (d, J = 8.0 Hz, 1H, Ar); ^{13}C-NMR (125 MHz, CDCl$_3$) δ: 21.0, 43.8, 44.9, 56.2, 112.2, 121.2, 123.0, 124.5, 125.3, 126.2, 128.7, 129.4, 129.5, 132.4, 141.7, 146.3, 153.2, 155.1; HRMS (FAB): m/z calcd for C$_{18}$H$_{17}$ClN$_3$OS [M + H]$^+$ 358.0781; found: 358.0777.

3.1.61 Synthesis of 9-(3-Chloro-6-methoxyphenyl)-3,4-dihydro-2H,6H-pyrimido[1,2-c][1,3]benzothiazin-6-imine (31x)

N-(*tert*-Butyl)-9-(3-chloro-6-methoxyphenyl)-3,4-dihydro-2H,6H-pyrimido [1,2-c][1,3]benzothiazin-6-imine (29x). Using the general procedure as described for **22l**, N-(*tert*-butyl)-9-bromo-3,4-dihydro-2H,6H-pyrimido[1,2-c][1,3]benzo-thiazin-6-imine **22k** (52.8 mg, 0.15 mmol) was allowed to react with 3-chloro-6-methoxyphenylboronic acid (33.6 mg, 0.18 mmol) for 1 h. Purification by flash chromatography over aluminum oxide with *n*-hexane–EtOAc (1:0 to 9:1) gave the title compound **29x** as a colorless oil (54.9 mg, 88 %): IR (neat) cm^{-1}: 1591 (C=N); ^{1}H-NMR (400 MHz, CDCl$_3$) δ: 1.39 (s, 9H, 3 × CH$_3$), 1.89–1.95 (m, 2H, CH$_2$), 3.64 (t, J = 5.5 Hz, 2H, CH$_2$), 3.78 (s, 3H, CH$_3$), 3.88 (t, J = 6.1 Hz, 2H, CH$_2$), 6.88 (d, J = 9.5 Hz, 1H, Ar), 7.26–7.34 (m, 4H, Ar), 8.21 (d, J = 8.3 Hz, 1H, Ar); ^{13}C-NMR (100 MHz, CDCl$_3$) δ: 21.9, 30.0 (3C), 45.1, 45.4, 54.1, 55.9, 112.6, 125.1, 125.8, 126.7, 127.2, 128.1, 128.8, 128.8, 130.2, 130.6, 138.4, 139.2, 147.8, 155.1; HRMS (FAB): m/z calcd for C$_{22}$H$_{25}$ClN$_3$OS [M + H]$^+$ 414.1407; found: 414.1410.

Compound 31x. Using the general procedure as described for **25a**, compound **29x** (33.2 mg, 0.080 mmol) was allowed to react for 3 h. Purification by flash chromatography over aluminum oxide with *n*-hexane–EtOAc (9:1 to 7:3) gave the title compound **31x** as colorless solid (16.3 mg, 57 %): mp 175–178 °C (from CHCl$_3$–*n*-hexane); IR (neat) cm^{-1}: 1620 (C=N), 1568 (C=N); ^{1}H-NMR (400 MHz, CDCl$_3$) δ: 1.96–2.02 (m, 2H, CH$_2$), 3.71 (t, $J = 5.6$ Hz, 2H, CH$_2$), 3.79 (s, 3H, CH$_3$), 4.04 (t, $J = 6.2$ Hz, 2H, CH$_2$), 6.89 (d, $J = 8.5$ Hz, 1H, Ar), 7.19 (d, $J = 1.7$ Hz, 1H, Ar), 7.26–7.30 (m, 2H, Ar), 7.36 (dd, $J = 8.3, 1.7$ Hz, 1H, Ar), 8.27 (d, $J = 8.3$ Hz, 1H, Ar); ^{13}C-NMR (125 MHz, CDCl$_3$) δ: 20.9, 43.8, 44.8, 55.9, 112.5, 124.2, 125.3, 125.8, 127.5, 128.6, 128.9, 130.1 (2C), 130.1, 139.8, 146.6, 153.3, 155.0; HRMS (FAB): *m/z* calcd for C$_{18}$H$_{17}$ClN$_3$OS [M + H]$^+$ 358.0781; found: 358.0783.

3.1.62 Synthesis of 3,4-Dihydro-9-(naphthalen-2-yl)-2H, 6H-pyrimido[1,2-c][1,3]benzothiazin-6-imine (32a)

N-(tert-Butyl)-3,4-dihydro-9-(naphthalen-2-yl)-2H,6H-pyrimido[1,2-c][1,3] benzothiazin-6-imine (30a). Using the general procedure as described for **22l**, N-(*tert*-butyl)-9-bromo-3,4-dihydro-2H,6H-pyrimido[1,2-c][1,3]benzothiazin-6-imine **22k** (52.8 mg, 0.15 mmol) was allowed to react with 2-naphthaleneboronic acid (30.9 mg, 0.18 mmol) for 1 h. Purification by flash chromatography over aluminum oxide with *n*-hexane–EtOAc (1:0 to 9:1) gave the title compound **30a** as colorless solid (56.3 mg, 94 %): mp 144–145 °C (from *n*-hexane); IR (neat) cm^{-1}: 1592 (C=N); ^{1}H-NMR (500 MHz, CDCl$_3$) δ: 1.42 (s, 9H, 3 × CH$_3$), 1.91–1.96 (m, 2H, CH$_2$), 3.65 (t, $J = 5.2$ Hz, 2H, CH$_2$), 3.90 (t, $J = 5.7$ Hz, 2H, CH$_2$), 7.45–7.51 (m, 3H, Ar), 7.55 (d, $J = 8.0$ Hz, 1H, Ar), 7.71 (d, $J = 8.0$ Hz, 1H, Ar), 7.84–7.91 (m, 3H, Ar), 8.04 (s, 1H, Ar), 8.29 (d, $J = 8.6$ Hz, 1H, Ar); ^{13}C-NMR (125 MHz, CDCl$_3$) δ: 21.9, 30.0 (3C), 45.1, 45.5, 54.2, 122.9, 125.0 (2C), 126.0, 126.3, 126.5, 126.6, 127.6, 128.3, 128.6, 129.0, 129.6, 132.9, 133.5, 136.7, 138.3, 142.8, 147.7; HRMS (FAB): *m/z* calcd for C$_{25}$H$_{26}$N$_3$S [M + H]$^+$ 400.1847; found: 400.1848.

Compound 32a. Using the general procedure as described for **25a**, compound **30a** (45.3 mg, 0.11 mmol) was allowed to react for 3 h. Purification by flash chromatography over aluminum oxide with *n*-hexane–EtOAc (9:1 to 7:3) gave the title compound **32a** as colorless solid (28.2 mg, 73 %): mp 143–145 °C (from CHCl$_3$–*n*-hexane); IR (neat) cm^{-1}: 1620 (C=N), 1568 (C=N); ^{1}H-NMR (400 MHz, CDCl$_3$) δ: 1.96-2.02 (m, 2H, CH$_2$), 3.72 (t, $J = 5.6$ Hz, 2H, CH$_2$), 4.04 (t, $J = 6.2$ Hz, 2H, CH$_2$), 7.36 (d, $J = 1.8$ Hz, 1H, Ar), 7.49–7.52 (m, 2H, Ar), 7.57 (dd, $J = 8.5, 1.8$ Hz, 1H, Ar), 7.68 (dd, $J = 8.5, 1.8$ Hz, 1H, Ar), 7.84–7.91 (m, 3H, Ar), 8.02 (d, $J = 1.5$ Hz, 1H, Ar), 8.33 (d, $J = 8.5$ Hz, 1H, Ar); ^{13}C-NMR (100 MHz, CDCl$_3$) δ: 21.0, 43.8, 44.9, 122.0, 124.9, 125.3 (2C), 126.1, 126.4, 126.5, 127.6, 128.3, 128.7, 129.4 (2C), 133.0, 133.5, 136.3, 143.4, 146.5, 153.3; HRMS (FAB): *m/z* calcd for C$_{21}$H$_{18}$N$_3$S [M + H]$^+$ 344.1221; found: 344.1222.

3.1.63 Synthesis of 3,4-Dihydro-9-(naphthalen-1-yl)-2H, 6H-pyrimido[1,2-c][1,3]benzothiazin-6-imine (32b)

N-(tert-Butyl)-3,4-dihydro-9-(naphthalen-1-yl)-2H,6H-pyrimido[1,2-c][1,3] benzothiazin-6-imine (30b). Using the general procedure as described for **22l**, N-(tert-butyl)-9-bromo-3,4-dihydro-2H,6H-pyrimido[1,2-c][1,3]benzothiazin-6-imine **22k** (52.8 mg, 0.15 mmol) was allowed to react with 1-naphthaleneboronic acid (30.9 mg, 0.18 mmol) for 1 h. Purification by flash chromatography over aluminum oxide with n-hexane–EtOAc (1:0 to 9:1) gave the title compound **30b** as a colorless oil (58.4 mg, 97 %): IR (neat) cm^{-1}: 1590 (C=N); ^{1}H-NMR (400 MHz, CDCl$_3$) δ: 1.40 (s, 9H, 3 × CH$_3$), 1.92–1.98 (m, 2H, CH$_2$), 3.67 (t, J = 5.6 Hz, 2H, CH$_2$), 3.92 (t, J = 6.1 Hz, 2H, CH$_2$), 7.25–7.25 (m, 1H, Ar), 7.34 (dd, J = 8.3, 1.7 Hz, 1H, Ar), 7.39–7.52 (m, 4H, Ar), 7.86 (d, J = 8.3 Hz, 2H, Ar), 7.89 (d, J = 7.6 Hz, 1H, Ar), 8.31 (d, J = 8.3 Hz, 1H, Ar); ^{13}C-NMR (100 MHz, CDCl$_3$) δ: 22.0, 30.0 (3C), 45.1, 45.4, 54.2, 125.3, 125.6, 125.6, 125.9, 126.3, 126.7, 126.8, 128.0, 128.2, 128.3, 128.3, 129.1, 131.2, 133.7, 138.2, 138.6, 142.8, 147.8; HRMS (FAB): m/z calcd for C$_{25}$H$_{26}$N$_3$S [M + H]$^+$ 400.1847; found: 400.1845.

Compound 32b. Using the general procedure as described for **25a**, compound **30b** (46.4 mg, 0.12 mmol) was allowed to react for 3 h. Purification by flash chromatography over aluminum oxide with n-hexane–EtOAc (9:1 to 7:3) gave the title compound **32b** as colorless solid (34.4 mg, 86 %): mp 146–148 °C (from CHCl$_3$–n-hexane); IR (neat) cm^{-1}: 1620 (C=N), 1568 (C=N); ^{1}H-NMR (400 MHz, CDCl$_3$) δ: 1.98–2.04 (m, 2H, CH$_2$), 3.73 (t, J = 5.6 Hz, 2H, CH$_2$), 4.06 (t, J = 6.2 Hz, 2H, CH$_2$), 7.17 (d, J = 1.7 Hz, 1H, Ar), 7.35–7.52 (m, 5H, Ar), 7.81–7.91 (m, 3H, Ar), 8.33 (d, J = 8.3 Hz, 1H, Ar); ^{13}C-NMR (100 MHz, CDCl$_3$) δ: 21.1, 43.8, 45.0, 124.7, 125.2, 125.4, 125.6, 126.0, 126.4, 126.8, 128.3, 128.4 (2C), 128.7, 128.9, 131.1, 133.7, 138.2, 143.4, 146.5, 153.3; HRMS (FAB): m/z calcd for C$_{21}$H$_{18}$N$_3$S [M + H]$^+$ 344.1221; found: 344.1221.

3.1.64 Synthesis of 3,4-Dihydro-9-(3,4-methylenedioxyphenyl)-2H,6H-pyrimido[1,2-c][1,3]benzothiazin-6-imine (32c)

N-(tert-Butyl)-3,4-dihydro-9-(3,4-methylenedioxyphenyl)-2H,6H-pyrimido [1, 2-c][1,3]benzothiazin-6-imine (30c). Using the general procedure as described for **22l**, N-(tert-butyl)-9-bromo-3,4-dihydro-2H,6H-pyrimido[1,2-c][1,3]benzothiazin-6-imine **22k** (52.8 mg, 0.15 mmol) was allowed to react with 3,4-(methylenedioxy)phenylboronic acid (29.9 mg, 0.18 mmol) for 1 h. Purification by flash chromatography over aluminum oxide with n-hexane–EtOAc (1:0 to 9:1) gave the title compound **30c** as colorless solid (54.6 mg, 93 %): mp 173 °C (from CHCl$_3$–n-hexane); IR (neat) cm^{-1}: 1591 (C=N); ^{1}H-NMR (400 MHz, CDCl$_3$) δ: 1.40 (s,

9H, 3 × CH$_3$), 1.89–1.95 (m, 2H, CH$_2$), 3.63 (t, J = 5.5 Hz, 2H, CH$_2$), 3.88 (t, J = 6.2 Hz, 2H, CH$_2$), 5.99 (s, 2H, CH$_2$), 6.87 (d, J = 8.8 Hz, 1H, Ar), 7.05–7.07 (m, 2H, Ar), 7.24 (d, J = 2.0 Hz, 1H, Ar), 7.34 (dd, J = 8.3, 2.0 Hz, 1H, Ar), 8.21 (d, J = 8.3 Hz, 1H, Ar); ^{13}C-NMR (100 MHz, CDCl$_3$) δ: 21.9, 30.0 (3C), 45.1, 45.4, 54.2, 101.3, 107.4, 108.6, 120.7, 122.3, 124.5, 126.2, 128.9, 129.5, 133.7, 138.3, 142.6, 147.7, 147.7, 148.3; HRMS (FAB): m/z calcd for C$_{22}$H$_{24}$N$_3$O$_2$S [M + H]$^+$ 394.1589; found: 394.1592.

Compound 32c. Using the general procedure as described for **25a**, compound **30c** (40.1 mg, 0.102 mmol) was allowed to react for 2 h. Purification by flash chromatography over aluminum oxide with n-hexane–EtOAc (9:1 to 7:3) gave the title compound **32c** as colorless solid (17.0 mg, 49 %): mp 169–170 °C (from CHCl$_3$–n-hexane); IR (neat) cm^{-1}: 1619 (C=N), 1568 (C=N); ^{1}H-NMR (400 MHz, CDCl$_3$) δ: 1.95–2.01 (m, 2H, CH$_2$), 3.70 (t, J = 5.5 Hz, 2H, CH$_2$), 4.03 (t, J = 6.1 Hz, 2H, CH$_2$), 6.00 (s, 2H, CH$_2$), 6.87 (d, J = 7.8 Hz, 1H, Ar), 7.04–7.06 (m, 2H, Ar), 7.15 (d, J = 1.7 Hz, 1H, Ar), 7.20 (br s, 1H, NH), 7.36 (dd, J = 8.3, 1.7 Hz, 1H, Ar), 8.24 (d, J = 8.3 Hz, 1H, Ar); ^{13}C-NMR (100 MHz, CDCl$_3$) δ: 21.0, 43.8, 45.0, 101.3, 107.4, 108.7, 120.8, 121.3, 124.8, 125.1, 129.3, 129.3, 133.4, 143.1, 146.4, 147.9, 148.3, 153.4; HRMS (FAB): m/z calcd for C$_{18}$H$_{16}$N$_3$O$_2$S [M + H]$^+$ 338.0963; found: 338.0960.

3.1.65 Synthesis of 3,4-Dihydro-9-(2,3-dihydrobenzo[b][1,4]dioxin-6-yl)-2H,6H-pyrimido[1,2-c][1,3]benzothiazin-6-imine (32d)

N-(*tert*-Butyl)-3,4-dihydro-9-(2,3-dihydrobenzo[b][1,4]dioxin-6-yl)-2H,6H-pyrimido[1,2-c][1,3]benzothiazin-6-imine (30d). Using the general procedure as described for **22l**, N-(*tert*-butyl)-9-bromo-3,4-dihydro-2H,6H-pyrimido[1,2-c][1,3]benzothiazin-6-imine **22k** (52.8 mg, 0.15 mmol) was allowed to react with (2,3-dihydrobenzo[b][1,4]dioxin-6-yl)boronic acid (32.3 mg, 0.18 mmol) for 1 h. Purification by flash chromatography over aluminum oxide with n-hexane–EtOAc (1:0 to 9:1) gave the title compound **30d** as a colorless oil (63.9 mg, 96 %): IR (neat) cm^{-1}: 1586 (C=N); ^{1}H-NMR (500 MHz, CDCl$_3$) δ: 1.40 (s, 9H, 3 × CH$_3$), 1.89–1.94 (m, 2H, CH$_2$), 3.63 (t, J = 5.4 Hz, 2H, CH$_2$), 3.88 (t, J = 6.0 Hz, 2H, CH$_2$), 4.28 (s, 4H, 2 × CH$_2$), 6.92 (d, J = 8.6 Hz, 1H, Ar), 7.07–7.09 (m, 1H, Ar), 7.11 (d, J = 2.3 Hz, 1H, Ar), 7.26 (s, 1H, Ar), 7.36 (t, J = 4.0 Hz, 1H, Ar), 8.21 (d, J = 8.6 Hz, 1H, Ar); ^{13}C-NMR (125 MHz, CDCl$_3$) δ: 21.9, 30.0 (3C), 45.1, 45.4, 54.1, 64.4, 64.5, 115.8, 117.6, 120.0, 122.2, 124.4, 126.1, 128.9, 129.4, 132.9, 138.4, 142.3, 143.8 (2C), 147.7; HRMS (FAB): m/z calcd for C$_{23}$H$_{26}$N$_3$O$_2$S [M + H]$^+$ 408.1746; found: 408.1746.

Compound 32d. Using the general procedure as described for **25a**, compound **30d** (45.1 mg, 0.11 mmol) was allowed to react for 3 h. Purification by flash chromatography over aluminum oxide with n-hexane–EtOAc (9:1 to 7:3) gave the title compound **32d** as colorless solid (26.7 mg, 69 %): mp 174–176 °C (from CHCl$_3$–*n*-hexane); IR (neat) cm^{-1}: 1619 (C=N), 1567 (C=N); ^{1}H-NMR (500 MHz, CDCl$_3$) δ: 1.95-2.00 (m, 2H, CH$_2$), 3.70 (t, J = 5.7 Hz, 2H, CH$_2$), 4.03 (t, J = 6.3 Hz, 2H, CH$_2$), 4.28 (s, 4H, 2 × CH$_2$), 6.92 (d, J = 8.6 Hz, 1H, Ar), 7.06 (dd, J = 8.6, 2.3 Hz, 1H, Ar), 7.09 (d, J = 2.3 Hz, 1H, Ar), 7.17 (d, J = 2.0 Hz, 1H, Ar), 7.19 (br s, 1H, NH), 7.38 (dd, J = 8.6, 2.0 Hz, 1H, Ar), 8.23 (d, J = 8.6 Hz, 1H, Ar; ^{13}C-NMR (125 MHz, CDCl$_3$) δ: 21.0, 43.8, 44.9, 64.3, 64.4, 115.8, 117.7, 120.0, 121.2, 124.6, 125.0, 129.2, 129.3, 132.5, 142.8, 143.8, 143.9, 146.4, 153.4; HRMS (FAB): *m/z* calcd for C$_{19}$H$_{18}$N$_3$O$_2$S [M + H]$^+$ 352.1120; found: 352.1121.

3.1.66 Synthesis of 3,4-Dihydro-9-(quinolin-6-yl)-2H, 6H-pyrimido[1,2-c][1,3]benzothiazin-6-imine (32e)

N-(*tert*-Butyl)-3,4-dihydro-9-(quinolin-6-yl)-2*H*,6*H*-pyrimido[1,2-*c*][1,3]benzo-thiazin-6-imine (30e). Using the general procedure as described for **22l**, *N*-(*tert*-butyl)-9-bromo-3,4-dihydro-2*H*,6*H*-pyrimido[1,2-*c*][1,3]benzothiazin-6-imine **22k** (52.8 mg, 0.15 mmol) was allowed to react with 6-quinolineboronic acid (31.1 mg, 0.18 mmol) for 1 h. Purification by flash chromatography over aluminum oxide with *n*-hexane–EtOAc (9:1 to 1:1) gave the title compound **30e** as colorless solid (36.2 mg, 60 %): mp 179–180 °C (from CHCl$_3$–*n*-hexane); IR (neat) cm^{-1}: 1590 (C=N); ^{1}H-NMR (400 MHz, CDCl$_3$) δ: 1.42 (s, 9H, 3 × CH$_3$), 1.91–1.97 (m, 2H, CH$_2$), 3.66 (t, J = 5.4 Hz, 2H, CH$_2$), 3.91 (t, J = 6.1 Hz, 2H, CH$_2$), 7.42–7.46 (m, 2H, Ar), 7.55 (dd, J = 8.3, 1.8 Hz, 1H, Ar), 7.95 (dd, J = 8.8, 2.0 Hz, 1H, Ar), 8.00 (d, J = 1.5 Hz, 1H, Ar), 8.16–8.21 (m, 2H, Ar), 8.32 (d, J = 8.3 Hz, 1H, Ar), 8.93 (dd, J = 4.1, 1.5 Hz, 1H, Ar); ^{13}C-NMR (100 MHz, CDCl$_3$) δ: 21.9, 30.0 (3C), 45.1, 45.4, 54.2, 121.6, 123.1, 125.0, 125.7, 126.9, 128.4, 128.7, 129.1, 129.8, 130.1, 136.2, 137.5, 138.0, 142.0, 147.6, 148.0, 150.7; HRMS (FAB): *m/z* calcd for C$_{24}$H$_{25}$N$_4$S [M + H]$^+$ 401.1800; found: 401.1802.

Compound 32e. Using the general procedure as described for **25a**, compound **30e** (27.0 mg, 0.067 mmol) was allowed to react for 3 h. Purification by flash chromatography over aluminum oxide with *n*-hexane–EtOAc (9:1 to 1:1) gave the title compound **32e** as colorless solid (19.3 mg, 84 %): mp 165–167 °C (from CHCl$_3$–*n*-hexane); IR (neat) cm^{-1}: 1620 (C=N), 1572 (C=N); ^{1}H-NMR (400 MHz, CDCl$_3$) δ: 1.98–2.04 (m, 2H, CH$_2$), 3.73 (t, J = 5.5 Hz, 2H, CH$_2$), 4.06 (t, J = 6.1 Hz, 2H, CH$_2$), 7.38 (d, J = 1.7 Hz, 1H, Ar), 7.45 (dd, J = 8.0, 4.3 Hz, 1H, Ar), 7.58 (dd, J = 8.3, 1.7 Hz, 1H, Ar), 7.94 (dd, J = 8.8, 2.2 Hz, 1H, Ar), 8.01 (d, J = 2.2 Hz, 1H, Ar), 8.18 (d, J = 8.8 Hz, 1H, Ar), 8.22 (dd, J = 8.0, 1.5 Hz, 1H, Ar), 8.35 (d, J = 8.3 Hz, 1H, Ar), 8.94 (dd, J = 4.3, 1.5 Hz, 1H, Ar); ^{13}C-NMR (100 MHz, CDCl$_3$) δ: 21.0, 43.8, 45.0, 121.7, 122.1, 125.3, 125.8, 125.9, 128.3, 128.6, 129.5, 129.6, 130.2, 136.3, 137.2, 142.5, 146.3, 148.0, 150.8,

153.2; HRMS (FAB): m/z calcd for $C_{20}H_{17}N_4S$ $[M + H]^+$ 345.1174; found: 345.1175.

3.1.67 Synthesis of 3,4-Dihydro-9-[3-(trifluoromethylcarbonyl)indol-6-yl]-2H, 6H-pyrimido[1,2-c][1,3]benzothiazin-6-imine (32f)

N-(*tert*-Butyl)-3,4-dihydro-9-(indol-6-yl)-2H,6H-pyrimido[1,2-c][1,3]benzo-thiazin-6-imine (30f). Using the general procedure as described for **22l**, N-(*tert*-butyl)-9-bromo-3,4-dihydro-2H,6H-pyrimido[1,2-c][1,3]benzothiazin-6-imine **22k** (52.8 mg, 0.15 mmol) was allowed to react with indol-6-ylboronic acid (29.0 mg, 0.18 mmol) for 1 h. Purification by flash chromatography over aluminum oxide with n-hexane–EtOAc (2:1) gave the title compound **30f** as colorless solid (56.0 mg, 96 %): mp 256 °C (from MeOH–CHCl$_3$–n-hexane); IR (neat) cm^{-1}: 1589 (C=N); ^{1}H-NMR (500 MHz, DMSO-d_6) δ: 1.39 (s, 9H, 3 × CH$_3$), 1.82–1.87 (m, 2H, CH$_2$), 3.55 (t, J = 5.4 Hz, 2H, CH$_2$), 3.82 (t, J = 6.0 Hz, 2H, CH$_2$), 6.46 (s, 1H, Ar), 7.37 (d, J = 8.3 Hz, 1H, Ar), 7.42 (t, J = 2.6 Hz, 1H, Ar), 7.58–7.63 (m, 3H, Ar), 7.72 (s, 1H, Ar), 8.21 (d, J = 8.3 Hz, 1H, Ar), 11.21 (br s, 1H, NH); ^{13}C-NMR (125 MHz, CDCl$_3$–CD$_3$OD) δ: 21.6, 29.7 (3C), 44.6, 45.4, 54.2, 109.6, 118.8, 120.7 (2C), 122.8, 125.1, 125.3, 125.5, 127.8, 128.5, 129.3, 132.7, 136.3, 138.4, 144.5, 149.3; HRMS (FAB): m/z calcd for $C_{23}H_{25}N_4S$ $[M + H]^+$ 389.1800; found: 389.1800.

Compound 32f. TFA (17 mL) was added to a mixture of **30f** (31.9 mg, 0.082 mmol) and MS4 Å (4.5 g, powder, activated by heating with Bunsen burner) in CHCl$_3$ (3.0 mL) and MeOH (5 drops). After being stirred under reflux for 8.5 h, the mixture was concentrated. To a mixture of this residue in CHCl$_3$ was added dropwise Et$_3$N at 0 °C to adjust pH to 8–9. The whole was extracted with EtOAc. The extract was washed with sat. NaHCO$_3$, brine, and dried over MgSO$_4$. After concentration, the residue was purified by preparative TLC over aluminum oxide with CHCl$_3$–MeOH (98:2) to give the title compound **32f** as pale yellow solid (17.8 mg, 51 %): mp 270 °C (decomp.) (from MeOH–CHCl$_3$–n-hexane); IR (neat) cm^{-1}: 1662 (C=O), 1568 (C=N); ^{1}H-NMR (500 MHz, DMSO-d_6) δ: 1.86–1.91 (m, 2H, CH$_2$), 3.61 (t, J = 5.4 Hz, 2H, CH$_2$), 3.94 (t, J = 6.0 Hz, 2H, CH$_2$), 7.61–7.62 (m, 2H, Ar), 7.71 (dd, J = 8.6, 1.7 Hz, 1H, Ar), 7.88 (s, 1H, Ar), 8.24–8.28 (m, 2H, Ar), 8.55 (d, J = 1.7 Hz, 1H, Ar), 8.75 (s, 1H, NH), 12.82 (br s, 1H, NH); ^{13}C-NMR (100 MHz, DMSO-d_6) δ: 20.7, 43.1, 44.4, 108.8, 111.1, 116.8 (q, J = 291.3 Hz), 121.5, 121.7, 122.6, 124.4, 124.8, 125.7, 129.1, 129.6, 134.6, 137.2, 138.5 (q, J = 4.7 Hz), 142.4, 145.1, 149.7, 173.9 (q, J = 33.9 Hz); HRMS (FAB): m/z calcd for $C_{21}H_{16}F_3N_4OS$ $[M + H]^+$ 429.0997; found: 429.1001.

3.1.68 Synthesis of 3,4-Dihydro-9-[3-(trifluoromethylcarbonyl)indol-5-yl]-2H, 6H-pyrimido[1,2-c][1,3]benzothiazin-6-imine (32g)

N-(tert-Butyl)-3,4-dihydro-9-(indol-5-yl)-2H,6H-pyrimido[1,2-c][1,3]benzo-thiazin-6-imine (30g). Using the general procedure as described for **22l**, *N*-(*tert*-butyl)-9-bromo-3,4-dihydro-2*H*,6*H*-pyrimido[1,2-c][1,3]benzothiazin-6-imine **22k** (52.8 mg, 0.15 mmol) was allowed to react with indol-5-ylboronic acid (29.0 mg, 0.18 mmol) for 1 h. Purification by flash chromatography over aluminum oxide with *n*-hexane–EtOAc (2:1) gave the title compound **30g** as colorless solid (59.2 mg, > 99 %): mp 232–233 °C (from MeOH–CHCl$_3$–*n*-hexane); IR (neat) cm^{-1}: 1583 (C=N); ^{1}H-NMR (400 MHz, CDCl$_3$–CD$_3$OD) δ: 1.41 (s, 9H, 3 × CH$_3$), 1.91–1.97 (m, 2H, CH$_2$), 3.63 (t, *J* = 5.4 Hz, 2H, CH$_2$), 3.90 (t, *J* = 6.1 Hz, 2H, CH$_2$), 6.58 (d, *J* = 3.0 Hz, 1H, Ar), 7.23 (d, *J* = 3.0 Hz, 1H, Ar), 7.41–7.42 (m, 3H, Ar), 7.50 (dd, *J* = 8.5, 0.9 Hz, 1H, Ar), 7.86 (s, 1H, Ar), 8.18 (d, *J* = 8.5 Hz, 1H, Ar); ^{13}C-NMR (100 MHz, CDCl$_3$–CD$_3$OD) δ: 21.8, 29.9 (3C), 44.9, 45.5, 54.2, 102.7, 111.4, 119.2, 121.2, 122.7, 125.1, 125.1, 125.3, 128.3, 128.6, 129.3, 131.1, 135.7, 138.6, 144.6, 148.8; HRMS (FAB): *m/z* calcd for C$_{23}$H$_{25}$N$_4$S [M + H]$^+$ 389.1800; found: 389.1800.

Compound 32g. TFA (17 mL) was added to a mixture of **30g** (28.5 mg, 0.073 mmol) and MS4Å (4.5 g, powder, activated by heating with Bunsen burner) in CHCl$_3$ (3.0 mL). After being stirred under reflux for 10 h, the mixture was concentrated. To a mixture of this residue in CHCl$_3$ was added dropwise Et$_3$N at 0 °C to adjust pH to 8–9. The whole was extracted with EtOAc. The extract was washed with sat. NaHCO$_3$, brine, and dried over MgSO$_4$. After concentration, the residue was purified by preparative TLC over aluminum oxide with CHCl$_3$–MeOH (98:2) to give the compound **32g** as pale yellow solid (28.9 mg, 92 %): mp 250 °C (decomp.) (from MeOH–CHCl$_3$–*n*-hexane); IR (neat) cm^{-1}: 1659 (C=O), 1613 (C=N), 1561 (C=N); ^{1}H-NMR (500 MHz, DMSO-*d$_6$*) δ: 1.86–1.91 (m, 2H, CH$_2$), 3.61 (t, *J* = 5.4 Hz, 2H, CH$_2$), 3.94 (t, *J* = 5.7 Hz, 2H, CH$_2$), 7.56–7.59 (m, 2H, Ar), 7.67–7.72 (m, 2H, Ar), 8.28 (d, *J* = 8.6 Hz, 1H, Ar), 8.45 (s, 1H, Ar), 8.56 (d, *J* = 1.7 Hz, 1H, Ar), 8.75 (br s, 1H, NH), 12.82 (br s, 1H, NH); ^{13}C-NMR (100 MHz, DMSO-*d$_6$*) δ: 20.7, 43.1, 44.4, 109.1, 113.6, 116.8 (q, *J* = 290.5 Hz), 119.2, 121.6, 123.7, 124.6, 124.6, 126.4, 129.1, 129.6, 133.9, 136.6, 138.4 (q, *J* = 4.7 Hz), 143.0, 145.2, 149.7, 173.9 (q, *J* = 33.7 Hz); HRMS (FAB): *m/z* calcd for C$_{21}$H$_{16}$F$_3$N$_4$OS [M + H]$^+$ 429.0997; found: 429.0991.

3.1.69 Synthesis of 3,4-Dihydro-9-(pyridin-3-yl)-2H, 6H-pyrimido[1,2-c][1,3]benzothiazin-6-imine (32h)

N-(tert-Butyl)-3,4-dihydro-9-(pyridin-3-yl)-2H,6H-pyrimido[1,2-c][1,3]benzo-thiazin-6-imine (30h). Using the general procedure as described for **22l**,

N-(*tert*-butyl)-9-bromo-3,4-dihydro-2*H*,6*H*-pyrimido[1,2-*c*][1,3]benzothiazin-6-imine **22k** (52.8 mg, 0.15 mmol) was allowed to react with 3-pyridineboronic acid (22.1 mg, 0.18 mmol) for 1 h. Purification by flash chromatography over aluminum oxide with *n*-hexane–EtOAc (7:3) gave the title compound **30h** as colorless solid (45.9 mg, 87 %): mp 143–144 °C (from CHCl$_3$–*n*-hexane); IR (neat) cm^{-1}: 1595 (C=N); ^{1}H-NMR (400 MHz, CDCl$_3$) δ: 1.40 (s, 9H, 3 × CH$_3$), 1.91–1.96 (m, 2H, CH$_2$), 3.65 (t, J = 5.6 Hz, 2H, CH$_2$), 3.89 (t, J = 6.2 Hz, 2H, CH$_2$), 7.33 (d, J = 1.7 Hz, 1H, Ar), 7.36 (dd, J = 7.8, 4.9 Hz, 1H, Ar), 7.41 (dd, J = 8.5, 1.7 Hz, 1H, Ar), 7.87 (ddd, J = 7.8, 2.2, 1.5 Hz, 1H, Ar), 8.30 (d, J = 8.5 Hz, 1H, Ar), 8.62 (dd, J = 4.9, 1.5 Hz, 1H, Ar), 8.85 (d, J = 2.2 Hz, 1H, Ar); ^{13}C-NMR (100 MHz, CDCl$_3$) δ: 21.8, 30.0 (3C), 45.1, 45.4, 54.2, 122.8, 123.5, 124.7, 127.3, 129.3, 130.0, 134.2, 135.0, 137.8, 139.6, 147.5, 148.1, 149.2; HRMS (FAB): *m/z* calcd for C$_{20}$H$_{23}$N$_4$S [M + H]$^+$ 351.1643; found: 351.1645.

Compound 32h. Using the general procedure as described for **25a**, compound **30h** (36.3 mg, 0.10 mmol) was allowed to react for 2 h with TFA (1.0 mL) and MS4Å (150 mg). Purification by flash chromatography over aluminum oxide with *n*-hexane–EtOAc (1:1) gave the title compound **32h** as colorless solid (24.8 mg, 81 %): mp 191–193 °C (from CHCl$_3$–*n*-hexane); IR (neat) cm^{-1}: 1616 (C=N), 1568 (C=N); ^{1}H-NMR (400 MHz, CDCl$_3$) δ: 1.97–2.03 (m, 2H, CH$_2$), 3.72 (t, J = 5.5 Hz, 2H, CH$_2$), 4.05 (t, J = 6.1 Hz, 2H, CH$_2$), 7.24 (d, J = 1.7 Hz, 1H, Ar), 7.38 (dd, J = 7.9, 4.8 Hz, 1H, Ar), 7.43 (dd, J = 8.3, 1.7 Hz, 1H, Ar), 7.86 (dt, J = 7.9, 1.8 Hz, 1H, Ar), 8.33 (d, J = 8.3 Hz, 1H, Ar), 8.63 (dd, J = 4.8, 1.8 Hz, 1H, Ar), 8.83 (d, J = 1.8 Hz, 1H, Ar); ^{13}C-NMR (100 MHz, CDCl$_3$) δ: 21.0, 43.8, 44.9, 121.8, 123.6, 124.9, 126.2, 129.7, 129.8, 134.2, 134.7, 140.1, 146.2, 148.1, 149.3, 152.9; HRMS (FAB): *m/z* calcd for C$_{16}$H$_{15}$N$_4$S [M + H]$^+$ 295.1017; found: 295.1013.

3.1.70 Synthesis of 3,4-Dihydro-9-(pyridin-4-yl)-2*H*, 6*H*-pyrimido[1,2-c][1,3]benzothiazin-6-imine (32i)

N-(*tert*-Butyl)-3,4-dihydro-9-(pyridin-4-yl)-2*H*,6*H*-pyrimido[1,2-*c*][1,3]benzo-thiazin-6-imine (30i). Using the general procedure as described for **22l**, *N*-(*tert*-butyl)-9-bromo-3,4-dihydro-2*H*,6*H*-pyrimido[1,2-*c*][1,3]benzothiazin-6-imine **22k** (52.8 mg, 0.15 mmol) was allowed to react with 4-pyridineboronic acid (22.1 mg, 0.18 mmol) for 1 h. Purification by flash chromatography over aluminum oxide with *n*-hexane–EtOAc (7:3) gave the title compound **30i** as colorless solid (27.6 mg, 52 %): mp 196–197 °C (from CHCl$_3$–*n*-hexane); IR (neat) cm^{-1}: 1593 (C=N); ^{1}H-NMR (500 MHz, CDCl$_3$) δ: 1.41 (s, 9H, 3 × CH$_3$), 1.91–1.96 (m, 2H, CH$_2$), 3.65 (t, J = 5.4 Hz, 2H, CH$_2$), 3.89 (t, J = 6.0 Hz, 2H, CH$_2$), 7.38 (s, 1H, Ar), 7.45 (d, J = 8.6 Hz, 1H, Ar), 7.49 (d, J = 5.0 Hz, 2H, Ar), 8.30 (d, J = 8.6 Hz, 1H, Ar), 8.67 (d, J = 5.0 Hz, 2H, Ar); ^{13}C-NMR (100 MHz, CDCl$_3$) δ: 21.8, 30.0 (3C), 45.1, 45.4, 54.2, 121.4 (2C), 122.8, 124.4, 128.1, 129.3, 130.1, 137.6, 139.8, 146.6, 147.4,

150.4 (2C); HRMS (FAB): m/z calcd for $C_{20}H_{23}N_4S$ $[M + H]^+$ 351.1643; found: 351.1644.

Compound 32i. Using the general procedure as described for **25a**, compound **30i** (23.4 mg, 0.067 mmol) was allowed to react for 2 h. Purification by flash chromatography over aluminum oxide with n-hexane–EtOAc (1:1 to 0:1) gave the title compound **32i** as colorless solid (15.1 mg, 77 %): mp 154–155 °C (from $CHCl_3$–n-hexane); IR (neat) cm^{-1}: 1620 (C=N), 1575 (C=N); ^{1}H-NMR (500 MHz, $CDCl_3$) δ: 1.97–2.02 (m, 2H, CH_2), 3.72 (t, $J = 5.4$ Hz, 2H, CH_2), 4.04 (t, $J = 6.3$ Hz, 2H, CH_2), 7.29 (s, 1H, Ar), 7.46–7.48 (m, 3H, Ar), 8.33 (d, $J = 8.6$ Hz, 1H, Ar), 8.68 (d, $J = 4.6$ Hz, 2H, Ar); ^{13}C-NMR (100 MHz, $CDCl_3$) δ: 21.0, 43.8, 45.0, 121.4 (2C), 121.9, 124.8, 127.1, 129.7, 129.9, 140.3, 146.1, 146.3, 150.4 (2C), 152.8; *Anal.* calcd for $C_{16}H_{14}N_4S$: C, 65.28; H, 4.79; N, 19.03. Found: C, 65.35; H, 4.63; N, 19.24.

3.1.71 Synthesis of 9-(Furan-2-yl)-3,4-dihydro-2H, 6H-pyrimido[1,2-c][1,3]benzothiazin-6-imine (32j)

N-(*tert*-Butyl)-9-(furan-2-yl)-3,4-dihydro-2*H*,6*H*-pyrimido[1,2-*c*][1,3]benzothiazin-6-imine (30j). Using the general procedure as described for **22l**, N-(*tert*-butyl)-9-bromo-3,4-dihydro-2*H*,6*H*-pyrimido[1,2-*c*][1,3]benzothiazin-6-imine **22k** (52.8 mg, 0.15 mmol) was allowed to react with 2-furanboronic acid (20.1 mg, 0.18 mmol) for 1 h. Purification by flash chromatography over aluminum oxide with n-hexane–EtOAc (1:0 to 9:1) gave the title compound **30j** as colorless solid (42.4 mg, 83 %): mp 128–130 °C (from n-hexane); IR (neat) cm^{-1}: 1596 (C=N); ^{1}H-NMR (400 MHz, $CDCl_3$) δ: 1.40 (s, 9H, 3 × CH_3), 1.89–1.94 (m, 2H, CH_2), 3.63 (t, $J = 5.5$ Hz, 2H, CH_2), 3.87 (t, $J = 6.1$ Hz, 2H, CH_2), 6.47–6.49 (m, 1H, Ar), 6.71 (d, $J = 3.4$ Hz, 1H, Ar), 7.42–7.48 (m, 3H, Ar), 8.20 (d, $J = 8.3$ Hz, 1H, Ar); ^{13}C-NMR (100 MHz, $CDCl_3$) δ: 21.9, 30.0 (3C), 45.1, 45.4, 54.2, 106.7, 111.9, 119.2, 121.4, 126.3, 128.8, 129.6, 132.3, 138.2, 142.8, 147.6, 152.5.; HRMS (FAB): m/z calcd for $C_{19}H_{22}N_3OS$ $[M + H]^+$ 340.1484; found: 340.1484.

Compound 32j. Using the general procedure as described for **25a**, compound **30j** (30.3 mg, 0.089 mmol) was allowed to react for 4 h with TFA (1.0 mL) and MS4Å (150 mg). Purification by flash chromatography over aluminum oxide with n-hexane–EtOAc (7:3) gave the title compound **32j** as colorless solid (18.3 mg, 73 %): mp 133 °C (from $CHCl_3$–n-hexane); IR (neat) cm^{-1}: 1620 (C=N), 1567 (C=N); ^{1}H-NMR (400 MHz, $CDCl_3$) δ: 1.95–2.01 (m, 2H, CH_2), 3.70 (t, $J = 5.6$ Hz, 2H, CH_2), 4.02 (t, $J = 6.1$ Hz, 2H, CH_2), 6.49 (dd, $J = 3.4, 1.7$ Hz, 1H, Ar), 6.73 (d, $J = 3.4$ Hz, 1H, Ar), 7.21 (br s, 1H, NH), 7.33 (d, $J = 1.7$ Hz, 1H, Ar), 7.47–7.49 (m, 2H, Ar), 8.22 (d, $J = 8.5$ Hz, 1H, Ar); ^{13}C-NMR (100 MHz, $CDCl_3$) δ: 21.0, 43.8, 45.0, 107.1, 112.0, 118.2, 121.6, 125.2, 129.3, 129.5, 132.7, 143.0, 146.3, 152.2, 153.3; *Anal.* calcd for $C_{15}H_{13}N_3OS$: C, 63.58; H, 4.62; N, 14.83. Found: C, 63.40; H, 4.46; N, 14.72.

3.1.72   Synthesis of 9-(Benzofuran-2-yl)-3,4-dihydro-2H, 6H-pyrimido[1,2-c][1,3]benzothiazin-6-imine (32k)

9-(Benzofuran-2-yl)-N-(*tert*-butyl)-3,4-dihydro-2H,6H-pyrimido[1,2-c][1,3] ben-zothiazin-6-imine (30k) Using the general procedure as described for **22l**, N-(*tert*-butyl)-9-bromo-3,4-dihydro-2H,6H-pyrimido[1,2-c][1,3]benzothiazin-6-imine   **22k** (52.8 mg, 0.15 mmol) was allowed to react with 2-benzofuranboronic acid (29.2 mg, 0.18 mmol) for 1 h. Purification by flash chromatography over aluminum oxide with *n*-hexane–EtOAc (1:0 to 9:1) gave the title compound **30k** as colorless solid (54.3 mg, 93 %): mp 211 °C (from CHCl$_3$–*n*-hexane); IR (neat) cm^{-1}: 1595 (C=N); ^{1}H-NMR (500 MHz, CDCl$_3$) δ: 1.41 (s, 9H, 3 × CH$_3$), 1.90–1.94 (m, 2H, CH$_2$), 3.64 (t, J = 5.4 Hz, 2H, CH$_2$), 3.88 (t, J = 6.0 Hz, 2H, CH$_2$), 7.06 (s, 1H, Ar), 7.23 (t, J = 7.4 Hz, 1H, Ar), 7.30 (t, J = 7.4 Hz, 1H, Ar), 7.50 (d, J = 7.4 Hz, 1H, Ar), 7.58 (d, J = 7.4 Hz, 1H, Ar), 7.62–7.64 (m, 2H, Ar), 8.25 (d, J = 8.0 Hz, 1H, Ar); ^{13}C-NMR (125 MHz, CDCl$_3$) δ: 21.9, 30.0 (3C), 45.2, 45.4, 54.2, 103.0, 111.2, 120.4, 121.2, 122.3, 123.1, 124.9, 127.3, 128.9 (2C), 129.8, 132.0, 138.0, 147.5, 154.2, 155.0; HRMS (FAB): *m/z* calcd for C$_{23}$H$_{24}$N$_3$OS [M + H]$^+$ 390.1640; found: 390.1645.

    Compound 32k. Using the general procedure as described for **25a**, compound **30k** (41.5 mg, 0.11 mmol) was allowed to react for 3 h. Purification by flash chromatography over aluminum oxide with *n*-hexane–EtOAc (9:1 to 7:3) gave the title compound **32k** as pale yellow solid (30.5 mg, 85 %): mp 189–191 °C (from CHCl$_3$–*n*-hexane); IR (neat) cm^{-1}: 1620 (C=N), 1565 (C=N); ^{1}H-NMR (500 MHz, CDCl$_3$) δ: 1.95–2.00 (m, 2H, CH$_2$), 3.70 (t, J = 5.4 Hz, 2H, CH$_2$), 4.02 (t, J = 6.3 Hz, 2H, CH$_2$), 7.07 (s, 1H, Ar), 7.22–7.25 (m, 1H, Ar), 7.31 (t, J = 7.2 Hz, 1H, Ar), 7.50–7.51 (m, 2H, Ar), 7.58 (d, J = 7.2 Hz, 1H, Ar), 7.65 (dd, J = 8.6, 1.1 Hz, 1H, Ar), 8.28 (d, J = 8.6 Hz, 1H, Ar); ^{13}C-NMR (125 MHz, CDCl$_3$) δ: 21.0, 43.8, 45.0, 103.4, 111.3, 119.4, 121.2, 122.6, 123.2, 125.1, 126.2, 128.8, 129.3, 129.6, 132.4, 146.2, 153.1, 153.9, 155.1; HRMS (FAB): *m/z* calcd for C$_{19}$H$_{16}$N$_3$OS [M + H]$^+$ 334.1014; found: 334.1017.

3.1.73   Synthesis of 3,4-Dihydro-9-(thiophen-3-yl)-2H, 6H-pyrimido[1,2-c][1,3]benzothiazin-6-imine (32l)

N-(*tert*-Butyl)-3,4-dihydro-9-(thiophen-3-yl)-2H,6H-pyrimido[1,2-c][1,3]ben-zothiazin-6-imine (30l). Using the general procedure as described for **22l**, N-(*tert*-butyl)-9-bromo-3,4-dihydro-2H,6H-pyrimido[1,2-c][1,3]benzothiazin-6-imine   **22k** (52.8 mg, 0.15 mmol) was allowed to react with 3-thiopheneboronic acid (23.0 mg, 0.18 mmol) for 1 h. Purification by flash chromatography over aluminum oxide with *n*-hexane–EtOAc (1:0 to 9:1) gave the title compound **30l** as colorless solid (54.0 mg, >99 %): mp 132–133 °C (from *n*-hexane), IR (neat) cm^{-1}: 1592 (C=N); ^{1}H-NMR (400 MHz, CDCl$_3$) δ: 1.40 (s, 9H, 3 × CH$_3$), 1.89–1.95 (m, 2H, CH$_2$),

3.63 (t, J = 5.6 Hz, 2H, CH$_2$), 3.88 (t, J = 6.2 Hz, 2H, CH$_2$), 7.32 (d, J = 2.0 Hz, 1H, Ar), 7.36–7.43 (m, 3H, Ar), 7.50 (dd, J = 2.7, 1.5 Hz, 1H, Ar), 8.21 (d, J = 8.5 Hz, 1H, Ar); ^{13}C-NMR (100 MHz, CDCl$_3$) δ: 21.9, 30.0 (3C), 45.1, 45.4, 54.2, 121.5, 121.9, 124.1, 126.0, 126.3, 126.6, 129.0, 129.6, 137.4, 138.2, 140.6, 147.7; HRMS (FAB): m/z calcd for C$_{19}$H$_{22}$N$_3$S$_2$ [M + H]$^+$ 356.1255; found: 356.1253.

   Compound 32l. Using the general procedure as described for **25a**, compound **30l** (37.6 mg, 0.11 mmol) was allowed to react for 2 h with TFA (1.0 mL) and MS4Å (150 mg). Purification by flash chromatography over aluminum oxide with *n*-hexane–EtOAc (4:1) gave the title compound **32l** as colorless solid (25.9 mg, 82 %): mp 120–121 °C (from CHCl$_3$–*n*-hexane); IR (neat) cm^{-1}: 1619 (C=N), 1569 (C=N); ^{1}H-NMR (400 MHz, CDCl$_3$) δ: 1.96–2.01 (m, 2H, CH$_2$), 3.70 (t, J = 5.4 Hz, 2H, CH$_2$), 4.03 (t, J = 6.1 Hz, 2H, CH$_2$), 7.23 (d, J = 1.1 Hz, 1H, Ar), 7.35–7.41 (m, 2H, Ar), 7.44 (dd, J = 8.3, 1.1 Hz, 1H, Ar), 7.51–7.52 (m, 1H, Ar), 8.24 (d, J = 8.3 Hz, 1H, Ar); ^{13}C-NMR (100 MHz, CDCl$_3$) δ: 21.0, 43.8, 44.9, 120.9, 121.8, 124.4, 125.0, 125.9, 126.7, 129.4 (2C), 138.0, 140.3, 146.5, 153.3; *Anal.* calcd for C$_{15}$H$_{13}$N$_3$S$_2$: C, 60.17; H, 4.38; N, 14.03. Found: C, 60.12; H, 4.11; N, 14.04.

3.1.74 Synthesis of 9-(Benzothiophen-2-yl)-3,4-dihydro-2H,6H-pyrimido[1,2-c][1,3]benzothiazin-6-imine (32m)

9-(Benzothiophen-2-yl)-*N*-(*tert*-butyl)-3,4-dihydro-2*H*,6*H*-pyrimido[1,2-*c*][1,3] benzothiazin-6-imine (30m). Using the general procedure as described for **22l**, *N*-(*tert*-butyl)-9-bromo-3,4-dihydro-2*H*,6*H*-pyrimido[1,2-*c*][1,3]benzothiazin-6-imine **22k** (52.8 mg, 0.15 mmol) was allowed to react with 2-benzothiopheneboronic acid (32.0 mg, 0.18 mmol) for 1 h. Purification by flash chromatography over aluminum oxide with *n*-hexane–EtOAc (1:0 to 9:1) gave the title compound **30m** as colorless solid (44.8 mg, 74 %): mp 222–223 °C (from CHCl$_3$–*n*-hexane); IR (neat) cm^{-1}: 1590 (C=N); ^{1}H-NMR (500 MHz, CDCl$_3$) δ: 1.41 (s, 9H, 3 × CH$_3$), 1.90–1.94 (m, 2H, CH$_2$), 3.63 (t, J = 5.4 Hz, 2H, CH$_2$), 3.88 (t, J = 6.3 Hz, 2H, CH$_2$), 7.30–7.36 (m, 2H, Ar), 7.42 (d, J = 1.7 Hz, 1H, Ar), 7.52 (dd, J = 8.6, 1.7 Hz, 1H, Ar), 7.58 (s, 1H, Ar), 7.76 (dd, J = 7.2, 1.4 Hz, 1H, Ar), 7.82 (d, J = 8.0 Hz, 1H, Ar), 8.23 (d, J = 8.6 Hz, 1H, Ar); ^{13}C-NMR (125 MHz, CDCl$_3$) δ: 21.8, 30.0 (3C), 45.1, 45.4, 54.2, 120.7, 121.9, 122.3, 123.8, 123.9, 124.7, 124.8, 127.2, 129.0, 129.9, 136.0, 137.9, 139.7, 140.4, 142.3, 147.5; HRMS (FAB): m/z calcd for C$_{23}$H$_{24}$N$_3$S$_2$ [M + H]$^+$ 406.1412; found: 406.1407.

   Compound 32m. Using the general procedure as described for **25a**, compound **30m** (35.8 mg, 0.09 mmol) was allowed to react for 3 h. Purification by flash chromatography over aluminum oxide with *n*-hexane–EtOAc (9:1 to 7:3) gave the title compound **32m** as pale yellow solid (26.9 mg, 86 %): mp 198–200 °C (from

MeOH–CHCl$_3$–n-hexane); IR (neat) cm^{-1}: 1615 (C=N), 1567 (C=N); ^{1}H-NMR (500 MHz, CDCl$_3$–CD$_3$OD) δ: 1.95–2.00 (m, 2H, CH$_2$), 3.69 (t, J = 5.4 Hz, 2H, CH$_2$), 4.01 (t, J = 6.0 Hz, 2H, CH$_2$), 7.25 (br s, 1H, NH), 7.32–7.37 (m, 3H, Ar), 7.54 (dd, J = 8.6, 1.7 Hz, 1H, Ar), 7.58 (s, 1H, Ar), 7.77 (t, J = 4.0 Hz, 1H, Ar), 7.82 (d, J = 7.4 Hz, 1H, Ar), 8.23 (d, J = 8.6 Hz, 1H, Ar); ^{13}C-NMR (125 MHz, CDCl$_3$–CD$_3$OD) δ: 20.9, 43.9, 44.9, 120.9, 121.0, 122.3, 123.9, 124.2, 124.7, 124.9, 125.9, 129.4, 129.6, 136.6, 139.7, 140.3, 141.8, 146.4, 153.2; HRMS (FAB): m/z calcd for C$_{19}$H$_{16}$N$_3$S$_2$ [M + H]$^+$ 350.0786; found: 350.0785.

3.1.75 Synthesis of 3,4-Dihydro-9-(1H-pyrazol-1-yl)-2H, 6H-pyrimido[1,2-c][1,3]benzothiazin-6-imine (32n)

N-(*tert*-Butyl)-3,4-dihydro-9-(1*H*-pyrazol-1-yl)-2*H*,6*H*-pyrimido[1,2-*c*][1,3] benzothiazin-6-imine (30n). To a solution of *N*-(*tert*-butyl)-9-bromo-3,4-dihydro-2*H*,6*H*-pyrimido[1,2-*c*][1,3]benzothiazin-6-imine **22k** (52.8 mg, 0.15 mmol), pyrazole (12.3 mg, 0.18 mmol), CuCl (1.5 mg, 0.015 mmol) and K$_2$CO$_3$ (21.8 mg, 0.16 mol) in *N*-methylpyrrolidone (0.3 mL) was added acetylacetone (3.8 µL, 0.038 mmol) under an Ar atmosphere. After being stirred at 130 °C for 19 h, EtOAc was added. The organic layers were washed with H$_2$O, and dried over MgSO$_4$. After concentration, the residue was purified by flash chromatography over aluminum oxide with *n*-hexane–EtOAc (7:3) to give the title compound **30n** as colorless solid (39.8 mg, 71 %): mp 132–133 °C (from CHCl$_3$–*n*-hexane); IR (neat) cm^{-1}: 1597 (C=N); ^{1}H-NMR (400 MHz, CDCl$_3$) δ: 1.39 (s, 9H, 3 × CH$_3$), 1.90–1.96 (m, 2H, CH$_2$), 3.63 (t, J = 5.6 Hz, 2H, CH$_2$), 3.88 (t, J = 6.2 Hz, 2H, CH$_2$), 6.48 (dd, J = 2.7, 1.8 Hz, 1H, Ar), 7.47 (dd, J = 8.8, 2.2 Hz, 1H, Ar), 7.56 (d, J = 2.2 Hz, 1H, Ar), 7.73 (d, J = 1.8 Hz, 1H, Ar), 7.94 (d, J = 2.7 Hz, 1H, Ar), 8.28 (d, J = 8.8 Hz, 1H, Ar); ^{13}C-NMR (100 MHz,CDCl$_3$) δ: 21.8, 30.0 (3C), 45.0, 45.4, 54.2, 108.2, 114.3, 115.9, 125.4, 126.7, 129.9, 130.8, 137.7, 141.0, 141.7, 147.3; HRMS (FAB): m/z calcd for C$_{18}$H$_{22}$N$_5$S [M + H]$^+$ 340.1596; found: 340.1598.

Compound 32n. Using the general procedure as described for **25a**, compound **30n** (21.6 mg, 0.064 mmol) was allowed to react for 1 h. Purification by preparative TLC over aluminum oxide with CHCl$_3$ gave the title compound **32n** as colorless solid (16.2 mg, 89 %): mp 158–159 °C (from CHCl$_3$–*n*-hexane); IR (neat) cm^{-1}: 1615 (C=N), 1561 (C=N); ^{1}H-NMR (400 MHz, CDCl$_3$) δ: 1.96–2.02 (m, 2H, CH$_2$), 3.70 (t, J = 5.6 Hz, 2H, CH$_2$), 4.03 (t, J = 6.2 Hz, 2H, CH$_2$), 6.49 (dd, J = 2.6, 1.8 Hz, 1H, Ar), 7.27 (s, 1H, NH), 7.48–7.51 (m, 2H, Ar), 7.74 (d, J = 1.5 Hz, 1H, Ar), 7.95 (d, J = 2.7 Hz, 1H, Ar), 8.31 (d, J = 9.5 Hz, 1H, Ar); ^{13}C-NMR (100 MHz, CDCl$_3$) δ: 21.0, 43.8, 44.9, 108.4, 113.3, 116.1, 124.3, 126.7, 130.3, 130.6, 141.3, 141.8, 145.9, 152.8; HRMS (FAB): m/z calcd for C$_{14}$H$_{14}$N$_5$S [M + H]$^+$ 284.0970; found: 284.0966.

3.1.76 Synthesis of 3,4-Dihydro-9-(1H-imidazol-1-yl)-2H,6H-pyrimido[1,2-c][1,3]benzothiazin-6-imine (32o)

N-(*tert*-Butyl)-3,4-dihydro-9-(1*H*-imidazol-1-yl)-2*H*,6*H*-pyrimido[1,2-c][1,3] benzothiazin-6-imine (30o). Using the general procedure as described for **30n**, N-(*tert*-butyl)-9-bromo-3,4-dihydro-2*H*,6*H*-pyrimido[1,2-c][1,3]benzothiazin-6-imine **22k** (52.8 mg, 0.15 mmol) was allowed to react with imidazole (12.3 mg, 0.18 mmol) for 3 h. Purification by flash chromatography over aluminum oxide with n-hexane–EtOAc (1:1 to 0:1) gave the title compound **30o** as colorless solid (25.8 mg, 51 %): mp 170–171 °C (from CHCl$_3$–n-hexane); IR (neat) cm^{-1}: 1599 (C=N); ^{1}H-NMR (400 MHz, CDCl$_3$) δ: 1.40 (s, 9H, 3 × CH$_3$), 1.91–1.96 (m, 2H, CH$_2$), 3.64 (t, J = 5.6 Hz, 2H, CH$_2$), 3.89 (t, J = 6.2 Hz, 2H, CH$_2$), 7.15 (d, J = 2.3 Hz, 1H, Ar), 7.22 (dd, J = 8.8, 2.3 Hz, 1H, Ar), 7.22 (s, 1H, Ar), 7.30 (s, 1H, Ar), 7.90 (s, 1H, Ar), 8.32 (d, J = 8.8 Hz, 1H, Ar); ^{13}C-NMR (100 MHz, CDCl$_3$) δ: 21.8, 30.0 (3C), 45.0, 45.5, 54.3, 116.2, 117.7, 118.4, 126.5, 130.5, 130.8, 131.3, 135.3, 136.9, 138.4, 146.9; HRMS (FAB): m/z calcd for C$_{18}$H$_{22}$N$_5$S [M + H]$^+$ 340.1596; found: 340.1598.

Compound 32o. Using the general procedure as described for **25a**, compound **30o** (20.6 mg, 0.061 mmol) was allowed to react for 1 h. Purification by preparative TLC over aluminum oxide with EtOAc–MeOH (9:1) gave the title compound **32o** as colorless solid (9.7 mg, 56 %): mp 183–185 °C (from CHCl$_3$–n-hexane); IR (neat) cm^{-1}: 1622 (C=N), 1561 (C=N); ^{1}H-NMR (400 MHz, CDCl$_3$) δ: 1.97-2.03 (m, 2H, CH$_2$), 3.71 (t, J = 5.7 Hz, 2H, CH$_2$), 4.04 (t, J = 6.1 Hz, 2H, CH$_2$), 7.07 (d, J = 2.2 Hz, 1H, Ar), 7.22 (s, 1H, Ar), 7.25 (dd, J = 8.8, 2.2 Hz, 1H, Ar), 7.29 (s, 1H, Ar), 7.89 (s, 1H, Ar), 8.36 (d, J = 8.8 Hz, 1H, Ar); ^{13}C-NMR (125 MHz, CDCl$_3$) δ: 20.9, 43.8, 44.9, 115.2, 117.7, 118.6, 125.5, 130.9, 131.0, 131.1, 135.3, 138.8, 145.6, 152.2; HRMS (FAB): m/z calcd for C$_{14}$H$_{14}$N$_5$S [M + H]$^+$ 284.0970; found: 284.0966.

3.1.77 Synthesis of 2,3-Dihydro-5H-imidazo[1,2-c][1,3] benzothiazin-5-imine (36a)

Using the general procedure as described for **25a**, N-(*tert*-butyl)-2,3-dihydroimidazo[1,2-c][1,3]benzothiazin-5-imine **35a** (18.4 mg, 0.07 mmol) was allowed to react for 12 h with TFA (1.0 mL) and MS4Å (150 mg). Purification by flash chromatography over silica gel with n-hexane–EtOAc (1:1) gave the title compound **36a** as colorless solid (11.1 mg, 78 %): mp 176–178 °C (from CHCl$_3$–n-hexane); IR (neat) cm^{-1}: 1621 (C=N), 1585 (C=N); ^{1}H-NMR (500 MHz, CDCl$_3$) δ: 4.11 (4H, s, 2 × CH$_2$), 5.82 (1H, br s, NH), 7.12 (1H, d, J = 8.0 Hz, Ar), 7.24-7.28 (1H, m, Ar), 7.38–7.42 (1H, m, Ar), 8.20 (1H, dd, J = 7.7, 1.4 Hz, Ar); ^{13}C-

NMR (125 MHz, CDCl$_3$) δ: 47.3, 52.9, 120.8, 123.8, 126.5, 129.1, 132.0, 132.4, 150.0, 154.0; HRMS (FAB): m/z calcd for C$_{10}$H$_{10}$N$_3$S [M + H]$^+$ 204.0595; found: 204.0600.

3.1.78 Synthesis of 6H,8H-Quinazolino[3,2-c][1,3] benzothiazin-6-imine (36b)

N-(tert-Butyl)-6H,8H-quinazolino[3,2-c][1,3]benzothiazin-6-imine (35b). To a solution of 2-fluorobenzaldehyde (1.41 g, 11.39 mmol) in t-BuOH (38 mL) was added 2-aminobenzylamine **33b** (1.53 g, 12.53 mmol). The mixture was stirred at 80 °C for 30 min, and then K$_2$CO$_3$ (4.73 g, 34.18 mmol) and I$_2$ (3.61 g, 14.24 mmol) were added. After being stirred at same temperature for 4 h, the mixture was quenched with sat. Na$_2$SO$_3$. The organic layer was separated and concentrated. The resulting solid was dissolved with H$_2$O and CHCl$_3$, and then pH was adjusted to 12–14 with 5 N NaOH. The whole was extracted with CHCl$_3$. The extract was washed with brine, dried over Na$_2$SO$_4$. To a mixture of resulting residue in DMAc (7.4 mL) were added KOt-Bu (496 mg, 4.42 mmol) and *tert*-butylisothiocyanate (0.56 mL, 4.42 mmol) under an N$_2$ atmosphere. After being stirred at 80 °C for 2.5 h, sat. NH$_4$Cl was added. The whole was extracted with EtOAc. The extract was washed with brine, and dried over Na$_2$SO$_4$. After concentration, the residue was purified by flash chromatography over silica gel with n-hexane–EtOAc (1:0 to 9:1) to give the title compound **35b** as yellow solid (114.1 mg, 3.1 % over 2 steps): mp 92.2 °C; IR (neat) cm^{-1}: 1588 (C=N); ^{1}H-NMR (300 MHz, CDCl$_3$) δ: 1.42 (9H, s, 3 × CH$_3$), 5.10 (2H, s, CH$_2$), 7.08–7.23 (3H, m, Ar), 7.27–7.40 (4H, m, Ar), 8.43 (1H, dd, J = 8.0, 1.4 Hz, Ar); ^{13}C-NMR (75 MHz, CDCl$_3$) δ: 29.9 (3C), 46.2, 54.7, 124.0, 124.8, 124.9, 125.4, 125.8, 126.4, 127.7, 128.3, 129.0, 129.4, 130.7, 138.3, 141.1, 148.3; *Anal.* calcd for C$_{19}$H$_{19}$N$_3$S: C, 70.99; H, 5.96; N, 13.07. Found: C, 71.05; H, 5.99; N, 12.91.

Compound 36b. TFA (0.5 mL) was added to **35b** (100 mg, 0.311 mmol). After being stirred under reflux for 30 min, the mixture was added dropwise to Et$_3$N at 0 °C to adjust pH to 8–9. The whole was extracted with EtOAc. The extract was washed with sat. NaHCO$_3$ aq., brine, and dried over Na$_2$SO$_4$. After concentration, the residue was purified by preparative TLC over aluminum oxide with n-hexane–EtOAc (9:1) to give the title compound 36b as colorless solid (13 mg, 16 %): mp 133–135 °C (from CHCl$_3$–n-hexane); IR (neat) cm^{-1}: 1594 (C=N), 1541 (C=N); ^{1}H-NMR (500 MHz, CDCl$_3$) δ: 5.27 (2H, s, CH$_2$), 7.08–7.13 (2H, m, Ar), 7.16 (1H, t, J = 7.2 Hz, Ar), 7.26–7.34 (3H, m, Ar), 7.39 (1H, td, J = 6.9, 1.1 Hz, Ar), 7.59 (1H, br s, NH), 8.50 (1H, d, J = 8.0 Hz, Ar); ^{13}C-NMR (125 MHz, CDCl$_3$) δ: 45.1, 123.0, 123.8, 125.4, 125.6, 126.3, 126.5, 126.6, 128.5, 129.2, 129.5, 131.1, 140.1, 146.3, 153.3; HRMS (FAB): m/z calcd for C$_{15}$H$_{12}$N$_3$S [M + H]$^+$ 266.0752; found: 266.0750.

3.1.79 Synthesis of (±)-3,4-Dihydro-3-methyl-2H, 6H-pyrimido[1,2-c][1,3]benzothiazin-6-imine (36c)

2-(2-Fluorophenyl)-5-methyl-1,4,5,6-tetrahydropyrimidine (34c). 2-Fluorobenz-aldehyde (0.62 g, 5.0 mmol) was subjected to the general procedure for **18j** using 2-methylpropylenediamine **33c** (0.48 g, 5.5 mmol) to give the title compound **34c** as colorless crystals (0.72 g, 75 %): mp 98–99 °C (from $CHCl_3$–n-hexane); IR (neat) cm^{-1}: 1628 (C=N); ^{1}H-NMR (500 MHz, $CDCl_3$) δ: 1.01 (3H, d, $J = 6.9$ Hz, CH_3), 1.92–1.99 (1H, m, CH), 3.06 (2H, dd, $J = 13.2$, 9.7 Hz, 2 × CH), 3.52 (2H, dd, $J = 13.2$, 3.4 Hz, 2 × CH), 5.27 (1H, br s, NH), 7.04 (1H, dd, $J = 11.7$, 8.3 Hz, Ar), 7.15 (1H, t, $J = 7.4$ Hz, Ar), 7.30–7.35 (1H, m, Ar), 7.81 (1H, td, $J = 7.4$, 1.7 Hz, Ar); ^{13}C-NMR (125 MHz, $CDCl_3$) δ: 16.8, 25.2, 49.4 (2C), 115.9 (d, $J = 24.0$ Hz), 124.2, 124.3 (d, $J = 3.6$ Hz), 130.6 (d, $J = 3.6$ Hz), 130.8 (d, $J = 8.4$ Hz), 151.3, 160.1 (d, $J = 247.1$ Hz); ^{19}F-NMR (500 MHz, $CDCl_3$) δ: -117.1; HRMS (FAB): m/z calcd for $C_{11}H_{14}FN_2$ [M + H]$^+$ 193.1141; found: 193.1136.

(±)-N-(-($tert$-Butyl)-3,4-dihydro-3-methyl-2H,6H-pyrimido[1,2-c][1,3]ben-zothiazin-6-imine (35c). Using the general procedure as described for **22e**, compound **34c** (384.5 mg, 2.0 mmol) was allowed to react at 80 °C for 2 h. Purification by flash chromatography over aluminum oxide with n-hexane–EtOAc (1:0 to 95:5) gave the title compound **35c** as colorless solid (288.4 mg, 50 %): mp 60–62 °C (from n-hexane); IR (neat) cm^{-1}: 1598 (C=N), 1570 (C=N); ^{13}C-NMR (500 MHz, $CDCl_3$) δ: 1.05 (3H, d, $J = 6.3$ Hz, CH_3), 1.39 (9H, s, 3 × CH_3), 1.91–1.99 (1H, m, CH), 3.09–3.17 (2H, m, CH_2), 3.72 (1H, dt, $J = 15.5$, 3.7 Hz, CH), 4.19 (1H, dt, $J = 13.7$, 3.7 Hz, CH), 7.11 (1H, d, $J = 8.0$ Hz, Ar), 7.19 (1H, t, $J = 8.0$ Hz, Ar), 7.30 (1H, t, $J = 8.0$ Hz, Ar), 8.19 (1H, d, $J = 8.0$ Hz, Ar); ^{13}C-NMR (125 MHz, $CDCl_3$) δ: 16.7, 26.9, 30.0 (3C), 51.6, 52.4, 54.2, 124.4, 126.0, 127.7, 128.5, 129.1, 130.1, 138.4, 147.6; HRMS (FAB): m/z calcd for $C_{16}H_{22}N_3S$ [M + H]$^+$ 288.1534; found: 288.1535.

Compound 36c. Using the general procedure as described for **25a**, compound **35c** (57.5 mg, 0.20 mmol) was allowed to react for 1 h. Purification by flash chromatography over aluminum oxide with n-hexane–EtOAc (9:1) gave the title compound **36c** as colorless solid (36.7 mg, 79 %): mp 82–84 °C (from $CHCl_3$–n-hexane); IR (neat) cm^{-1}: 1621 (C=N), 1574 (C=N); ^{1}H-NMR (500 MHz, $CDCl_3$) δ: 1.09 (3H, d, $J = 6.3$ Hz, CH_3), 1.96–2.08 (1H, m, CH), 3.19 (1H, dd, $J = 15.8$, 10.6 Hz, CH), 3.27 (1H, dd, $J = 13.0$, 10.6 Hz, CH), 3.80 (1H, ddd, $J = 15.8$, 4.5, 3.2 Hz, CH), 4.37 (1H, ddd, $J = 13.0$, 4.5, 3.2 Hz, CH), 7.04 (1H, d, $J = 7.4$ Hz, Ar), 7.18–7.25 (2H, m, Ar, NH), 7.33 (1H, td, $J = 7.4$, 1.4 Hz, Ar), 8.23 (1H, dd, $J = 8.3$, 1.4 Hz, Ar); ^{13}C-NMR (125 MHz, $CDCl_3$) δ: 16.4, 26.1, 49.9, 52.2, 123.5, 126.3, 126.6, 128.8, 128.9, 130.6, 146.2, 153.4; *Anal.* calcd for $C_{12}H_{13}N_3S$: C, 62.31; H, 5.66; N, 18.17. Found: C, 62.04; H, 5.75; N, 17.88.

3.1.80 Synthesis of 3,4-Dihydro-3,3-dimethyl-2H, 6H-pyrimido[1,2-c][1,3]benzothiazin-6-imine (36d)

2-(2-Fluorophenyl)-5,5-dimethyl-1,4,5,6-tetrahydropyrimidine (34d). 2-Fluo-robenzaldehyde (0.62 g, 5.0 mmol) was subjected to the general procedure for **18j** using 2,2-dimethylpropylenediamine **33d** (0.56 g, 5.5 mmol) to give the title compound 34d as colorless crystals (0.82 g, 79 %): mp 150–153 °C (from $CHCl_3$–n-hexane); IR (neat) cm^{-1}: 1629 (C=N); ^{1}H-NMR (400 MHz, $CDCl_3$) δ: 1.02 (6H, s, 2 × CH_3), 3.13 (4H, s, 2 × CH_2), 5.14 (1H, br s, NH), 7.05 (1H, ddd, J = 11.7, 7.8, 1.0 Hz, Ar), 7.15 (1H, td, J = 7.8, 1.0 Hz, Ar), 7.30–7.35 (1H, m, Ar), 7.81 (1H, td, J = 7.8, 2.0 Hz, Ar); ^{13}C-NMR (100 MHz, $CDCl_3$) δ: 25.0 (2C), 26.2, 54.3 (2C), 115.8 (d, J = 23.2 Hz), 124.2, 124.3 (d, J = 3.3 Hz), 130.6 (d, J = 4.1 Hz), 130.8 (d, J = 9.1 Hz), 150.5 (d, J = 1.7 Hz), 160.2 (d, J = 247.5 Hz); ^{19}F-NMR (500 MHz, $CDCl_3$) δ: -117.3; HRMS (FAB): m/z calcd for $C_{12}H_{16}FN_2$ [M + H]$^+$ 207.1298; found: 207.1299.

N-(tert-Butyl)-3,4-dihydro-3,3-dimethyl-2H,6H-pyrimido[1,2-c][1,3]benzo-thiazin-6-imine (35d). Using the general procedure as described for **22e**, compound **34d** (412.5 mg, 2.0 mmol) was allowed to react at 80 °C for 2 h. Purification by flash chromatography over aluminum oxide with n-hexane–EtOAc (1:0 to 9:1) gave the title compound 35d as colorless solid (236.6 mg, 39 %): mp 70–72 °C (from n-hexane); IR (neat) cm^{-1}: 1602 (C=N), 1570 (C=N); ^{1}H-NMR (500 MHz, $CDCl_3$) δ: 1.01 (6H, s, 2 × CH_3), 1.39 (9H, s, 3 × CH_3), 3.33 (2H, s, CH_2), 3.58 (2H, s, CH_2), 7.12 (1H, d, J = 8.0 Hz, Ar), 7.20 (1H, t, J = 8.0 Hz, Ar), 7.31 (1H, td, J = 8.0, 1.1 Hz, Ar), 8.21 (1H, dd, J = 8.0, 1.1 Hz, Ar); ^{13}C-NMR (125 MHz, $CDCl_3$) δ: 24.8 (2C), 28.5, 29.9 (3C), 54.2, 55.7, 57.4, 124.5, 126.0, 127.5, 128.5, 129.1, 130.1, 138.7, 146.7; HRMS (FAB): m/z calcd for $C_{17}H_{24}N_3S$ [M + H]$^+$ 302.1691; found: 302.1695.

Compound 36d. Using the general procedure as described for **25a**, compound **35d** (60.3 mg, 0.20 mmol) was allowed to react for 1 h. Purification by flash chromatography over aluminum oxide with n-hexane–EtOAc (9:1) gave the title compound **36d** as colorless solid (42.0 mg, 86 %): mp 113–114 °C (from $CHCl_3$–n-hexane); IR (neat) cm^{-1}: 1627 (C=N), 1575 (C=N); ^{1}H-NMR (500 MHz, $CDCl_3$) δ: 1.05 (6H, s, 2 × CH_3), 3.41 (2H, s, CH_2), 3.74 (2H, s, CH_2), 7.05 (1H, dd, J = 7.6, 1.1 Hz, Ar), 7.21–7.25 (2H, m, Ar, NH), 7.34 (1H, td, J = 7.6, 1.4 Hz, Ar), 8.26 (1H, dd, J = 8.3, 1.4 Hz, Ar); ^{13}C-NMR (125 MHz, $CDCl_3$) δ: 24.6 (2C), 27.9, 54.0, 57.2, 123.5, 126.3, 126.3, 128.8, 128.9, 130.6, 145.3, 153.8; HRMS (FAB): m/z calcd for $C_{13}H_{16}N_3S$ [M + H]$^+$ 246.1065; found: 246.1069.

3.1.81 Synthesis of 2,3,4,5-Tetrahydro-7H-1,3-diazepino [1,2-c][1,3]benzothiazin-7-imine (36e)

N-(tert-Butoxycarbonyl)-2-(2-fluorophenyl)-4,5,6,7-tetrahydro-1,3-diazepine (37). To a solution of 2-fluorobenzaldehyde (2.48 g, 20.0 mmol) in t-BuOH (188 mL) was added 1,4-diaminobutane **33e** (2.21 mL, 22.0 mmol). The mixture was stirred at 70 °C for 30 min, and then K_2CO_3 (8.29 g, 60.0 mmol) and I_2 (6.35 g, 25 mmol) were added. After being stirred at same temperature for 3 h, the mixture was quenched with sat. Na_2SO_3. The organic layer was separated and concentrated. The resulting solid was dissolved with H_2O, and then pH was adjusted to 12–14 with 2 N NaOH. The whole was extracted with $CHCl_3$, and dried over Na_2SO_4. After concentration, Et_3N (8.67 mL, 60.0 mmol) and Boc_2O (13.8 mL, 60.0 mmol) were added to the solution of residue in CH_2Cl_2 (100 mL). After being stirred for 30 min at rt, sat. $NaHCO_3$ was added. After being stirred at rt for 1 h, the whole was extracted with $CHCl_3$. The extract was washed with brine, and dried over $MgSO_4$. After concentration, the residue was purified by column chromatography over silica gel with n-hexane–EtOAc (4:1) to give the title compound **37** as colorless solid (2.18 g, 37 %): mp 63–65 °C (from n-hexane); IR (neat) cm^{-1}: 1710 (C=O), 1631 (C=N); ^{1}H-NMR (500 MHz, CDCl$_3$) δ: 1.14 (9H, s, 3 × CH$_3$), 1.66–1.70 (2H, m, CH$_2$), 1.78–1.83 (2H, m, CH$_2$), 3.61 (2H, br s, CH$_2$), 3.76 (2H, t, J = 5.2 Hz, CH$_2$), 7.03 (1H, dd, J = 11.2, 8.3 Hz, Ar), 7.15 (1H, td, J = 7.7, 1.1 Hz, Ar), 7.33–7.38 (1H, m, Ar), 7.60 (1H, t, J = 7.7 Hz, Ar); ^{13}C-NMR (125 MHz, CDCl$_3$) δ: 23.2, 26.4, 27.7 (3C), 44.9, 50.7, 81.1, 115.7 (d, J = 21.6 Hz), 124.0 (d, J = 2.4 Hz), 126.5, 130.9 (d, J = 2.4 Hz), 131.1 (d, J = 8.4 Hz), 152.8, 154.8, 160.5 (d, J = 250.7 Hz); ^{19}F-NMR (500 MHz, CDCl$_3$) δ: −118.9; HRMS (FAB) m/z calcd for $C_{16}H_{22}FN_2O_2$ [M + H]$^+$ 293.1665; found: 293.1669.

2-(2-Fluorophenyl)-4,5,6,7-tetrahydro-1H-1,3-diazepine (34e). To a solution of **37** (877.1 mg, 3.0 mmol) in CH_2Cl_2 (6.0 mL) was added TFA (6.0 mL). The mixture was stirred under reflux for 2 h, mixture was washed with 2 N NaOH. The organic phase was dried over $MgSO_4$. After concentration, the residue was re-crystallized from $CHCl_3$–n-hexane to give the title compound **34e** as colorless crystals (461.2 mg, 80 %): mp 92 °C; IR (neat) cm^{-1}: 1627 (C=N); ^{1}H-NMR (500 MHz, CDCl$_3$) δ: 1.80–1.83 (4H, m, 2 × CH$_2$), 3.48 (4H, br s, 2 × CH$_2$), 4.86 (1H, br s, NH), 7.02–7.06 (1H, m, Ar), 7.12 (1H, td, J = 7.7, 1.1 Hz, Ar), 7.30–7.34 (1H, m, Ar), 7.63 (1H, td, J = 7.7, 1.7 Hz, Ar); ^{13}C-NMR (125 MHz, CDCl$_3$) δ: 28.4 (2C), 47.9 (2C), 115.7 (d, J = 22.8 Hz), 124.2 (d, J = 3.6 Hz), 127.0 (d, J = 12.0 Hz), 130.9 (d, J = 8.4 Hz), 131.2 (d, J = 3.6 Hz), 157.2, 160.4 (d, J = 247.1 Hz); ^{19}F-NMR (500 MHz, CDCl$_3$) δ: −117.7; HRMS (FAB) m/z calcd for $C_{11}H_{14}FN_2$ [M + H]$^+$ 193.1141; found: 193.1140.

N-(tert-Butyl)-7H-2,3,4,5-tetrahydro-1,3-diazepino[1,2-c][1,3]benzothiazin-7-imine (35e). Using the general procedure as described for **25e**, compound **34e** (192.2 mg, 1.0 mmol) was allowed to react at rt overnight. Purification by flash chromatography over silica gel with n-hexane–EtOAc (4:1) gave the title compound

35e as a yellow oil (50.3 mg, 18 %): IR (neat) cm^{-1}: 1588 (C=N); ^{1}H-NMR (400 MHz, CDCl$_3$) δ: 1.37 (9H, s, 3 × CH$_3$), 1.87–1.93 (4H, m, 2 × CH$_2$), 3.82 (2H, t, J = 5.4 Hz, CH$_2$), 3.88 (2H, t, J = 5.4 Hz, CH$_2$), 7.16–7.23 (2H, m, Ar), 7.26–7.31 (1H, m, Ar), 7.84 (1H, d, J = 7.1 Hz, Ar); ^{13}C-NMR (100 MHz, CDCl$_3$) δ: 23.3, 24.5, 30.2 (3C), 48.3, 49.2, 53.8, 124.9, 126.3, 127.0, 129.4, 129.7, 133.5, 140.0, 152.2; HRMS (FAB) m/z calcd for C$_{16}$H$_{22}$N$_3$S [M + H]$^+$ 288.1534; found: 288.1540.

Compound 36e. Using the general procedure as described for **25a**, compound **35e** (50.3 mg, 0.18 mmol) was allowed to react for 2 h. Purification by flash chromatography over aluminum oxide with n-hexane–EtOAc (4:1 to 2:1) gave the title compound 36e as a colorless oil (11.3 mg, 27 %): IR (neat) cm^{-1}: 1638 (C=N), 1578 (C=N); ^{1}H-NMR (400 MHz, CDCl$_3$) δ: 1.94–2.01 (4H, m, 2 × CH$_2$), 3.92 (2H, t, J = 5.5 Hz, CH$_2$), 3.96 (2H, t, J = 5.6 Hz, CH$_2$), 7.00 (1H, br s, NH), 7.12–7.14 (1H, m, Ar), 7.23–7.28 (1H, m, Ar), 7.31–7.35 (1H, m, Ar), 7.90 (1H, dd, J = 7.8, 1.5 Hz, Ar); ^{13}C-NMR (125 MHz, CDCl$_3$) δ: 23.3, 24.3, 47.5, 49.0, 124.2, 126.6, 127.5, 129.3, 129.9, 132.0, 151.0, 155.5; HRMS (FAB) m/z calcd for C$_{12}$H$_{14}$N$_3$S [M + H]$^+$ 232.0908; found: 232.0906.

3.1.82 Synthesis of 9-Bromo-2H-spiro(benzo[e]pyrimido [1,2-c][1,3]thiazine-3,1'-cyclohexan)-6(4H)-imine (49)

Cyclohexane-1,1-dicarbonitrile (42). To a solution of malononitrile (660.6 mg, 10.0 mmol) in DMF (25.0 mL) was added DBU (2.99 mL, 20.0 mmol). After being stirred at 50 °C for 2 h, a solution of 1,5-dibromopentane **38** (1.35 mL, 10.0 mmol) in DMF (10.0 mL) was added to the reaction mixture. After being stirred at the same temperature for additional 5 h, EtOAc was added. The mixture was washed with 5 % aq. NaHCO$_3$, and dried over MgSO$_4$. After concentration, the residue was purified by flash chromatography over silica gel with n-hexane–EtOAc (3:1). The resulting solid was recrystalized from CHCl$_3$–n-hexane to give the title compound 42 as colorless crystals (801.8 mg, 60 %): mp 62 °C, IR (neat) cm^{-1}: 2254 (C ≡ N); ^{1}H-NMR (400 MHz, CDCl$_3$) δ: 1.51–1.57 (m, 2H, CH$_2$), 1.73–1.78 (m, 4H, 2 × CH$_2$), 2.13 (t, J = 5.9 Hz, 4H, 2 × CH$_2$); ^{13}C-NMR (100 MHz, CDCl$_3$) δ: 21.6 (2C), 23.9, 32.4, 34.6 (2C), 115.9 (2C); MS (FAB) m/z (%): 135 (MH$^+$, 100).

3-(4-Bromo-2-fluorophenyl)-2,4-diazaspiro[5.5]undec-2-ene (45). To a solution of **42** (134.2 mg, 1.0 mmol) in THF (2.5 mL) was added BH$_3$-THF in THF (5.0 mL, 5.0 mmol, 1.0 M) at 0 °C under an Ar atmosphere. The mixture was warmed to rt. After being stirred at 65 °C for 5 h, the reaction mixture was cooled to 0 °C, and was added 1 N HCl. After being stirred at rt for 1 h, the mixture was basified with 2 N NaOH. The whole was extracted with CHCl$_3$ and dried over MgSO$_4$. After concentration, the residue was dissolved in t-BuOH (10.0 mL), and 4-bromo-2-fluorobenzaldehyde (203.0 mg, 1.0 mmol) was added.

After being stirred at 70 °C for 30 min, K_2CO_3 (414.6 mg, 3.0 mmol) and I_2 (317.3 mg, 1.25 mmol) were added. After being stirred at same temperature for 3 h, the reaction mixture was quenched with sat. Na_2SO_3 until the iodine color almost disappeared. The reaction mixture was basified with 2 N NaOH. The whole was extracted with $CHCl_3$, and dried over $MgSO_4$. After concentration, the residue was purified by flash chromatography over aluminum oxide with EtOAc–MeOH (1:0 to 95:5) gave the title compound **45** as colorless solid (200.1 mg, 62 %): mp 204–205 °C (from $CHCl_3$–n-hexane), IR (neat) cm^{-1}: 1626 (C=N); ^{1}H-NMR (500 MHz, $CDCl_3$) δ: 1.35–1.37 (m, 4H, 2 × CH_2), 1.47–1.49 (m, 6H, 3 × CH_2), 3.20 (s, 4H, 2 × CH_2), 5.07 (s, 1H, NH), 7.23 (dd, $J = 11.2, 2.0$ Hz, 1H, Ar), 7.28 (dd, $J = 8.3, 2.0$ Hz, 1H, Ar), 7.69 (t, $J = 8.3$ Hz, 1H, Ar); ^{13}C-NMR (125 MHz, $CDCl_3$) δ: 21.7 (2C), 26.5, 28.8 (2C), 33.5, 52.2 (2C), 119.4 (d, $J = 27.6$ Hz), 123.1 (d, $J = 12.0$ Hz), 123.5 (d, $J = 10.8$ Hz), 127.7 (d, $J = 3.6$ Hz), 131.8 (d, $J = 3.6$ Hz), 149.9, 159.8 (d, $J = 251.9$ Hz); ^{19}F-NMR (500 MHz, $CDCl_3$) δ: −114.6; HRMS (FAB): m/z calcd for $C_{15}H_{19}BrFN_2$ [M + H]$^+$ 325.0716; found: 325.0724.

9-Bromo-N-($tert$-butyl)-2H-spiro(benzo[e]pyrimido[1,2-c][1,3]thiazine-3,1′-cyclohexan)-6(4H)-imine (48). To a mixture of compound **45** (164.5 mg, 0.51 mmol) and NaH (40.8 mg, 1.02 mmol; 60 % oil suspension) in DMF (3.3 mL) was added t-BuNCS (129.4 μL, 1.02 mmol) under an Ar atmosphere. After being stirred at rt overnight, the reaction mixture was warmed at 60 °C. After being stirred at this temperature for 1 h, EtOAc was added. The resulting solution was washed with sat. $NaHCO_3$, brine, and dried over $MgSO_4$. After concentration, the residue was purified by flash chromatography over aluminum oxide with n-hexane–EtOAc (1:0 to 9:1) gave the title compound **48** as colorless solid (180.1 mg, 84 %): mp 118–119 °C (from n-hexane); IR (neat) cm^{-1}: 1578 (C=N); ^{1}H-NMR (500 MHz, $CDCl_3$) δ: 1.32–1.37 (m, 4H, 2 × CH_2), 1.38 (s, 9H, 3 × CH_3), 1.43–1.52 (m, 6H, 3 × CH_2), 3.38 (s, 2H, CH_2), 3.71 (s, 2H, CH_2), 7.26–7.31 (m, 2H, Ar), 8.04 (d, $J = 8.6$ Hz, 1H, Ar); ^{13}C-NMR (125 MHz, $CDCl_3$) δ: 21.8 (2C), 26.5, 29.9 (3C), 31.2, 33.4 (2C), 52.5, 54.2, 55.9, 124.3, 126.5, 126.8, 129.1, 130.0, 130.9, 137.6, 146.2; HRMS (FAB): m/z calcd for $C_{20}H_{27}BrN_3S$ [M + H]$^+$ 420.1109; found: 420.1117.

Compound 49. Using the general procedure as described for **25a**, **48** (124.7 mg, 0.3 mmol) was allowed to react under reflux for 2.5 h with TFA (3.0 mL) and MS4Å (0.45 g). Purification by flash chromatography over aluminum oxide with n-hexane–EtOAc (7:3) gave the title compound **49** as pale yellow solid (89.8 mg, 82 %): mp 130 °C (from $CHCl_3$–n-hexane); IR (neat) cm^{-1}: 1626 (C=N), 1572 (C=N); ^{1}H-NMR (400 MHz, $CDCl_3$) δ: 1.37–1.55 (m, 10H, 5 × CH_2), 3.46 (s, 2H, CH_2), 3.82 (s, 2H, CH_2), 7.21 (d, $J = 2.0$ Hz, 1H, Ar), 7.24 (s, 1H, NH), 7.33 (dd, $J = 8.8, 2.0$ Hz, 1H, Ar), 8.10 (d, $J = 8.8$ Hz, 1H, Ar); ^{13}C-NMR (100 MHz, $CDCl_3$) δ: 21.6 (2C), 26.3, 30.6, 33.3 (2C), 51.4, 55.4, 125.0, 125.3, 125.9, 129.5, 130.5, 130.7, 144.8, 152.7; HRMS (FAB): m/z calcd for $C_{16}H_{19}BrN_3S$ [M + H]$^+$ 364.0483; found: 364.0485.

3.1.83 Synthesis of 9-Bromo-2′,3′,5′,6′-tetrahydro-2H-spiro(benzo[e]pyrimido[1,2-c][1,3]thiazine-3,4′-pyran)-6(4H)-imine (51)

Dihydro-2H-pyran-4,4(3H)-dicarbonitrile (43). To a solution of malononitrile (660.6 mg, 10.0 mmol) in DMF (25.0 mL) was added DBU (2.99 mL, 20.0 mmol). After stirring at 50 °C for 2 h, the reaction mixture was added a solution of bis(2-chloroethyl)ether **39** (1.18 mL, 10.0 mmol) in DMF (10.0 mL). After being stirred at same temperature for 5 h, EtOAc was added. The mixture was washed with 5 % aq. NaHCO$_3$, and dried over MgSO$_4$. The filtrate was concentrated. The residue was purified by flash chromatography over silica gel with n-hexane–EtOAc (3:1). The resulting solid was recrystalized from CHCl$_3$–n-hexane to give the title compound **43** as colorless crystals (112.2 mg, 8 %): mp 96 °C, IR (neat) cm^{-1}: 2253 (C $\equiv$ N); ^{1}H-NMR (400 MHz, CDCl$_3$) δ: 2.24 (t, J = 5.2 Hz, 4H, 2 $\times$ CH$_2$), 3.87 (t, J = 5.2 Hz, 4H, 2 $\times$ CH$_2$); ^{13}C-NMR (100 MHz, CDCl$_3$) δ: 30.2, 33.8 (2C), 63.0 (2C), 114.9 (2C); MS (FAB) m/z (%): 137 (MH$^+$, 100).

3-(4-Bromo-2-fluorophenyl)-9-oxa-2,4-diazaspiro[5.5]undec-2-ene   (46). Using the general procedure as described for **45**, **43** (84.1 mg, 0.62 mmol) was allowed to react. Purification by flash chromatography over aluminum oxide with EtOAc–MeOH (1:0 to 95:5) gave the title compound **46** as colorless solid (21.4 mg, 11 %): mp 200–201 °C (from CHCl$_3$–n-hexane); IR (neat) cm^{-1}: 1626 (C=N); ^{1}H-NMR (500 MHz, CDCl$_3$) δ: 1.52 (t, J = 5.4 Hz, 4H, 2 $\times$ CH$_2$), 3.31 (s, 4H, 2 $\times$ CH$_2$), 3.72 (t, J = 5.4 Hz, 4H, 2 $\times$ CH$_2$), 4.03 (br s, 1H, NH), 7.25 (dd, J = 11.7, 2.0 Hz, 1H, Ar), 7.31 (dd, J = 8.6, 2.0 Hz, 1H, Ar), 7.70 (t, J = 8.6 Hz, 1H, Ar); ^{13}C-NMR (125 MHz, CDCl$_3$) δ: 27.0, 33.3 (2C), 51.4 (2C), 63.7 (2C), 119.5 (d, J = 27.6 Hz), 122.5 (d, J = 13.2 Hz), 124.0 (d, J = 9.6 Hz), 127.9 (d, J = 3.6 Hz), 131.7 (d, J = 3.6 Hz), 150.4, 159.7 (d, J = 251.9 Hz); HRMS (FAB): m/z calcd for C$_{14}$H$_{17}$BrFN$_2$O [M + H]$^+$ 327.0508; found: 327.0512.

9-Bromo-N-(tert-butyl)-2′,3′,5′,6′-tetrahydro-2H-spiro(benzo[e]pyrimido[1,2-c][1,3]thiazine-3,4′-pyran)-6(4H)-imine (50). Using the general procedure as described for **48**, **46** (21.4 mg, 0.065 mmol) was allowed to react at rt overnight. Purification by flash chromatography over aluminum oxide with n-hexane–EtOAc (1:0 to 9:1) gave the title compound 50 as colorless solid (21.5 mg, 78 %): mp 148–149 °C (from n-hexane); IR (neat) cm^{-1}: 1578 (C=N); ^{1}H-NMR (400 MHz, CDCl$_3$) δ: 1.38 (s, 9H, 3 $\times$ CH$_3$), 1.48–1.53 (m, 4H, 2 $\times$ CH$_2$), 3.48 (s, 2H, CH$_2$), 3.71–3.74 (m, 4H, 2 $\times$ CH$_2$), 3.84 (s, 2H, CH$_2$), 7.29–7.33 (m, 1H, Ar), 7.37 (d, J = 4.1 Hz, 1H, Ar), 8.05 (d, J = 8.5 Hz, 1H, Ar); ^{13}C-NMR (125 MHz, CDCl$_3$–CD$_3$OD) δ: 29.4, 29.9 (3C), 33.1 (2C), 51.4, 54.3, 55.5, 63.8 (2C), 124.6, 126.1, 126.9, 129.3, 130.0, 130.9, 137.6, 146.4; HRMS (FAB): m/z calcd for C$_{19}$H$_{25}$BrN$_3$OS [M + H]$^+$ 422.0902; found:422.0898.

Compound 51. Using the general procedure as described for **25a**, **50** (21.4 mg, 0.065 mmol) was allowed to react under reflux for 2.5 h with TFA (1.0 mL) and

MS4Å (150 mg). Purification by flash chromatography over aluminum oxide with *n*-hexane–EtOAc (7:3) gave the title compound **51** as colorless solid (12.3 mg, 66 %): mp 212–214 °C (from CHCl$_3$–*n*-hexane); IR (neat) cm^{-1}: 1626 (C=N), 1573 (C=N); ^{1}H-NMR (400 MHz, CDCl$_3$) δ: 1.54 (t, J = 5.4 Hz, 4H, 2 × CH$_2$), 3.56 (s, 2H, CH$_2$), 3.74 (t, J = 5.4 Hz, 4H, 2 × CH$_2$), 3.93 (s, 2H, CH$_2$), 7.22 (d, J = 2.0 Hz, 1H, Ar), 7.31 (br s, 1H, NH), 7.34 (dd, J = 8.8, 2.0 Hz, 1H, Ar), 8.10 (d, J = 8.8 Hz, 1H, Ar); ^{13}C-NMR (100 MHz, CDCl$_3$–CD$_3$OD) δ: 28.6, 32.9 (2C), 50.7, 54.6, 63.4 (2C), 124.7, 125.3, 125.9, 129.6, 130.3, 130.5, 145.4, 152.9; HRMS (FAB): *m/z* calcd for C$_{15}$H$_{17}$BrN$_3$OS [M + H]$^+$ 366.0276; found: 366.0280.

3.1.84 Synthesis of 9-Bromo-1′-(4-methoxybenzyl)-2H-spiro(benzo[e]pyrimido[1,2-c][1,3]thiazine-3,4′-piperidin)-6(4H)-imine (53a)

Bis(2-chloroethyl)-*N*-(4-methoxybenzyl)amine (41). To a suspension of bis(2-chloroethyl)amine hydrochloride **40** (8.92 g, 50.0 mmol) in CH$_2$Cl$_2$ (300 mL) were added Et$_3$N (2.89 mL, 100.0 mmol) and 4-methoxybenzoyl chloride (6.77 mL, 50.0 mmol). After being stirred at rt for 2 h, the reaction mixture was washed with 1 N HCl, sat. NaHCO$_3$, brine, and dried over MgSO$_4$. After concentration, the residue was dissolved in anhydrous Et$_2$O (250 mL), and LiAlH$_4$ (2.1 g, 55.0 mmol) was slowly added to the mixture at 0 °C under an Ar atmosphere. After being stirred at rt overnight, the reaction mixture was quenched by the addition of water, 2 N NaOH, and water. The mixture was dried over MgSO$_4$. After concentration, the residue was purified by flash chromatography over silica gel with *n*-hexane–EtOAc (19:1) to give the title compound **41** as colorless oil (9.88 g, 75 %): ^{1}H-NMR (400 MHz, CDCl$_3$) δ: 2.90 (t, J = 7.1 Hz, 4H, 2 × CH$_2$), 3.48 (t, J = 7.1 Hz, 4H, 2 × CH$_2$), 3.67 (s, 2H, CH$_2$), 3.80 (s, 3H, CH$_3$), 6.86 (d, J = 8.5 Hz, 2H, Ar), 7.24 (d, J = 8.5 Hz, 2H, Ar); ^{13}C-NMR (100 MHz, CDCl$_3$) δ: 42.0 (2C), 55.2, 56.2 (2C), 58.6, 113.8 (2C), 129.7 (2C), 130.7, 158.9; MS (FAB) *m/z* (%): 262 (MH$^+$, 100).

1-(4-Methoxybenzyl)piperidine-4,4-dicarbonitrile (44). To a solution of malononitrile (2.49 g, 37.7 mmol) in DMF (94.3 mL) was added K$_2$CO$_3$ (5.73 mg, 41.5 mmol). After being stirred at 65 °C for 2 h, a solution of **41** (9.88 mg, 37.7 mmol) in DMF (37.7 mL) was added. After being stirred at same temperature for 5 h, EtOAc was added. The mixture was washed with 5 % aq. NaHCO$_3$, and dried over MgSO$_4$. After concentration, the residue was purified by flash chromatography over silica gel with *n*-hexane–EtOAc (2:1) to give the title compound **44** as yellow oil (8.13 g, 85 %): IR (neat) cm^{-1}: 2248 (C $\equiv$ N); ^{1}H-NMR (400 MHz, CDCl$_3$) δ: 2.22 (t, J = 5.4 Hz, 4H, 2 × CH$_2$), 2.61 (br s, 4H, 2 × CH$_2$), 3.48 (s, 2H, CH$_2$), 3.80 (s, 3H, CH$_2$), 6.86 (d, J = 8.5 Hz, 2H, Ar), 7.19 (d, J = 8.8 Hz, 2H, Ar); ^{13}C-NMR (100 MHz, CDCl$_3$) δ: 31.1, 34.1 (2C), 48.5

(2C), 55.2, 61.9, 113.8 (2C), 115.4 (2C), 129.2, 130.1 (2C), 159.0; HRMS (FAB): m/z calcd for $C_{15}H_{18}N_3O$ [M + H]$^+$ 256.1450; found: 256.1454.

3-(4-Bromo-2-fluorophenyl)-9-(4-methoxybenzyl)-2,4,9-triazaspiro[5.5] undec-2-ene (47). Using the general procedure as described for **45**, **44** (4.05 g, 15.9 mmol) was allowed to react. Purification by flash chromatography over aluminum oxide with EtOAc–MeOH (1:0 to 95:5) to give the title compound **47** as colorless solid (752.6 mg, 11 %): mp 179–181 °C (from CHCl$_3$–n-hexane), IR (neat) cm^{-1}: 1630 (C=N); ^{1}H-NMR (500 MHz, CDCl$_3$) δ: 1.45 (t, J = 5.4 Hz, 4H, 2 × CH$_2$), 2.35 (t, J = 5.4 Hz, 4H, 2 × CH$_2$), 3.16 (s, 4H, 2 × CH$_2$), 3.40 (s, 2H, CH$_2$), 3.73 (s, 3H, CH$_3$), 4.63 (s, 1H, NH), 6.78 (d, J = 8.6 Hz, 2H, Ar), 7.14–7.23 (m, 4H, Ar), 7.62 (t, J = 8.3 Hz, 1H, Ar); ^{13}C-NMR (100 MHz, CDCl$_3$) δ: 27.3, 32.8 (2C), 49.1 (2C), 51.4 (2C), 55.2, 62.7, 113.5 (2C), 119.4 (d, J = 27.3 Hz), 122.7 (d, J = 12.4 Hz), 123.7 (d, J = 9.9 Hz), 127.8 (d, J = 3.3 Hz), 130.2, 130.3 (2C), 131.7 (d, J = 4.1 Hz), 150.3 (d, J = 1.7 Hz), 158.6, 159.7 (d, J = 251.6 Hz); ^{19}F-NMR (500 MHz, CDCl$_3$) δ: −114.6; HRMS (FAB): m/z calcd for $C_{22}H_{26}BrFN_3O$ [M + H]$^+$ 446.1243; found: 446.1237.

9-Bromo-N-(tert-butyl)-1′-(4-methoxybenzyl)-2H-spiro(benzo[e]pyrimido[1, 2-c][1,3]thiazine-3,4′-piperidin)-6(4H)-imine (52a). Using the general procedure as described for **48**, **47** (2.0 g, 4.48 mmol) was allowed to react at rt overnight. Purification by flash chromatography over aluminum oxide with n-hexane–EtOAc (1:0 to 9:1) to give the title compound **52a** as colorless solid (2.28 g, 94 %): mp 89–91 °C (from CHCl$_3$–n-hexane); IR (neat) cm^{-1}: 1577 (C=N); ^{1}H-NMR (500 MHz, CDCl$_3$) δ: 1.37 (s, 9H, 3 × CH$_3$), 1.49–1.52 (m, 4H, 2 × CH$_2$), 2.40–2.46 (m, 4H, 2 × CH$_2$), 3.41 (s, 2H, CH$_2$), 3.47 (s, 2H, CH$_2$), 3.75 (s, 2H, CH$_2$), 3.80 (s, 3H, CH$_3$), 6.85 (d, J = 8.6 Hz, 2H, Ar), 7.22 (d, J = 8.6 Hz, 2H, Ar), 7.28–7.31 (m, 2H, Ar), 8.03 (d, J = 8.6 Hz, 1H, Ar); ^{13}C-NMR (100 MHz, CDCl$_3$) δ: 29.7, 29.9 (3C), 32.6 (2C), 49.2 (2C), 51.6, 54.3, 55.2, 55.5, 62.7, 113.6 (2C), 124.5, 126.3, 126.8, 129.2, 130.0, 130.1, 130.4 (2C), 130.9, 137.5, 146.3, 158.7; HRMS (FAB): m/z calcd for $C_{27}H_{34}BrN_4OS$ [M + H]$^+$ 541.1637; found: 541.1633.

Compound 53a. Using the general procedure as described for **25a**, compound **52a** (448.1 mg, 0.83 mmol) was allowed to react for 2 h with TFA (10.0 mL) and MS4Å (1.50 g). Purification by flash chromatography over aluminum oxide with n-hexane–EtOAc (7:3) gave the title compound **53a** as colorless solid (288.8 mg, 72 %): mp 160–162 °C (from CHCl$_3$–n-hexane); IR (neat) cm^{-1}: 1626 (C=N), 1573 (C=N); ^{1}H-NMR (400 MHz, CDCl$_3$) δ: 1.54 (t, J = 5.5 Hz, 4H, 2 × CH$_2$), 2.39-2.51 (m, 4H, 2 × CH$_2$), 3.46 (s, 2H, CH$_2$), 3.48 (s, 2H, CH$_2$), 3.79 (s, 3H, CH$_3$), 3.86 (s, 2H, CH$_2$), 6.84 (d, J = 8.8 Hz, 2H, Ar), 7.20–7.22 (m, 3H, Ar), 7.28 (s, 1H, NH), 7.32 (dd, J = 8.8, 2.0 Hz, 1H, Ar), 8.08 (d, J = 8.8 Hz, 1H, Ar); ^{13}C-NMR (100 MHz, CDCl$_3$) δ: 29.2, 32.6 (2C), 49.1 (2C), 50.8, 55.1, 55.2, 62.6, 113.6 (2C), 125.1, 125.1, 125.9, 129.5, 130.3 (2C), 130.3, 130.4, 130.7, 145.0, 152.6, 158.7; HRMS (FAB): m/z calcd for $C_{23}H_{26}BrN_4OS$ [M + H]$^+$ 485.1011; found: 485.1010.

3.1.85 Synthesis of 9-Bromo-1'-(methoxycarbonyl)-2H-spiro(benzo[e]pyrimido[1,2-c][1,3]thiazine-3,4'-piperidin)-6(4H)-imine (53b)

9-Bromo-N-(*tert*-butyl)-1'-(methoxycarbonyl)-2*H*-spiro(benzo[e]pyrimido[1,2-c][1,3]thiazine-3,4'-piperidin)-6(4*H*)-imine (52b). To the solution of 9-bromo-N -(*tert*-butyl)-1'-(4-methoxybenzyl)-2H-spiro(benzo[e]pyrimido[1,2-c][1,3]thiazine-3,4'-piperidin)-6(4H)-imine **52a** (40.6 mg, 0.075 mmol) in CH$_2$Cl$_2$ (0.38 mL) was added methyl chloroformate (86.4 µL, 1.13 mmol) at 0 °C under an Ar atmosphere. After being stirred at same temperature for 30 min, the reaction mixture was concentrated. The residue was purified by flash chromatography over silica gel with *n*-hexane–EtOAc (1:1) to give the title compound **52b** as a colorless solid (29.2 mg, 81 %): mp 157–158 °C (from *n*-hexane); IR (neat) cm^{-1}: 1699 (C=O), 1577 (C=N); ^{1}H-NMR (400 MHz, CDCl$_3$) δ: 1.37 (s, 9H, 3 × CH$_3$), 1.46 (t, J = 5.6 Hz, 4H, 2 × CH$_2$), 3.44 (br s, 4H, 2 × CH$_2$), 3.56 (br s, 2H, CH$_2$), 3.70 (s, 3H, CH$_3$), 3.81 (s, 2H, CH$_2$), 7.29–7.33 (m, 2H, Ar), 8.05 (d, J = 8.5 Hz, 1H, Ar); ^{13}C-NMR (100 MHz, CDCl$_3$) δ: 29.9 (3C), 30.1, 32.2 (2C), 39.9 (2C), 50.8, 52.5, 54.3, 55.2, 124.7, 126.1, 126.8, 129.3, 130.0, 130.9, 137.7, 146.3, 155.9; HRMS (FAB): *m/z* calcd for C$_{21}$H$_{28}$BrN$_4$O$_2$S [M + H]$^+$ 479.1116; found: 479.1115.

Compound 53b. Using the general procedure as described for **25a**, compound **52b** (6.4 mg, 0.013 mmol) was allowed to react for 2 h with TFA (1.0 mL) and MS4Å (150 mg). Purification by flash chromatography over aluminum oxide with *n*-hexane–EtOAc (9:3) gave the title compound **53b** as colorless solid (4.0 mg, 73 %): mp 139–141 °C (from MeCN–H$_2$O); IR (neat) cm^{-1}: 1692 (C=O), 1626 (C=N), 1573 (C=N); ^{1}H-NMR (400 MHz, CDCl$_3$) δ: 1.49 (t, J = 5.7 Hz, 4H, 2 × CH$_2$), 3.45-3.57 (m, 6H, 3 × CH$_2$), 3.70 (s, 3H, CH$_3$), 3.91 (s, 2H, CH$_2$), 7.22 (d, J = 2.0 Hz, 1H, Ar), 7.31 (br s, 1H, NH), 7.34 (dd, J = 8.8, 2.0 Hz, 1H, Ar), 8.10 (d, J = 8.8 Hz, 1H, Ar); ^{13}C-NMR (100 MHz, CDCl$_3$) δ: 29.5, 32.2 (2C), 39.7 (2C), 50.0, 52.6, 54.6, 125.0, 125.3, 126.0, 129.6, 130.5, 130.6, 145.1, 152.6, 155.9; HRMS (FAB): *m/z* calcd for C$_{17}$H$_{20}$BrN$_4$O$_2$S [M + H]$^+$ 423.0490; found: 423.0492.

3.1.86 Synthesis of 1'-Acetyl-9-bromo-2H-spiro(benzo[e]pyrimido[1,2-c][1,3]thiazine-3,4'-piperidin)-6(4H)-imine (53c)

1'-Acetyl-9-bromo-N-(*tert*-butyl)-2*H*-spiro(benzo[e]pyrimido[1,2-c][1,3]thiazine-3,4'-piperidin)-6(4*H*)-imine (52c). Using the general procedure as described for **52b**, 9-bromo-N-(tert-butyl)-1'-(4-methoxybenzyl)-2H-spiro(benzo[e]pyrimido[1,2-c][1,3]thiazine-3,4'-piperidin)-6(4H)-imine **52a** (40.6 mg, 0.075 mmol) was allowed to react for 10 min with AcCl (53.3 μL, 0.75 mmol). Purification by flash chromatography over aluminum oxide with n-hexane–EtOAc (1:1) gave the title compound **52c** as colorless solid (33.3 mg, 96 %): mp 181–182 °C (from CHCl$_3$–n-hexane); IR (neat) cm^{-1}: 1632 (C=O), 1578 (C=N); ^{1}H-NMR (400 MHz, CDCl$_3$) δ: 1.37 (s, 9H, 3 × CH$_3$), 1.46–1.52 (m, 4H, 2 × CH$_2$), 2.10 (s, 3H, CH$_3$), 3.45–3.56 (m, 5H, 5 × CH), 3.72-3.78 (m, 2H, 2 × CH), 3.90 (d, J = 13.4 Hz, 1H, CH), 7.29–7.33 (m, 2H), 8.05 (d, J = 8.5 Hz, 1H); ^{13}C-NMR (100 MHz, CDCl$_3$) δ: 21.4, 29.9 (3C), 30.3, 32.2, 32.8, 37.5, 42.6, 50.8, 54.4, 55.1, 124.7, 126.0, 126.9, 129.3, 130.0, 130.8, 137.7, 146.3, 168.8; HRMS (FAB): m/z calcd for C$_{21}$H$_{28}$BrN$_4$OS [M + H]$^+$ 463.1167; found: 463.1164.

Compound 53c. Using the general procedure as described for **25a**, compound **52c** (6.5 mg, 0.014 mmol) was allowed to react for 2 h with TFA (1.0 mL) and MS4 Å (150 mg). Purification by flash chromatography over aluminum oxide with n-hexane–EtOAc (1:2) gave the title compound **53c** as colorless solid (4.5 mg, 79 %): mp 147–148 °C (from CHCl$_3$–n-hexane); IR (neat) cm^{-1}: 1625 (C=O), 1573 (C=N); ^{1}H-NMR (400 MHz, CDCl$_3$) δ: 1.49–1.55 (m, 4H, 2 × CH$_2$), 2.09 (s, 3H, CH$_3$), 3.47–3.61 (m, 5H, 5 × CH), 3.70–3.77 (m, 1H, CH), 3.84 (d, J = 13.4 Hz, 1H, CH), 4.04 (d, J = 13.2 Hz, 1H, CH), 7.22 (d, J = 1.2 Hz, 1H, Ar), 7.33–7.36 (m, 2H, Ar, NH), 8.10 (d, J = 8.8 Hz, 1H, Ar); ^{13}C-NMR (100 MHz, CDCl$_3$) δ: 21.4, 29.7, 32.1, 32.7, 37.3, 42.4, 49.8, 54.7, 124.9, 125.3, 126.0, 129.6, 130.4, 130.6, 145.1, 152.6, 168.9; Anal. calcd for C$_{17}$H$_{19}$BrN$_4$OS: C, 50.13; H, 4.70; N, 13.75. Found: C, 50.24; H, 4.78; N, 13.57.

3.1.87 Synthesis of 9-Bromo-1′-(methanesulfonyl)-2H-spiro(benzo[e]pyrimido[1,2-c][1,3]thiazine-3,4′-piperidin)-6(4H)-imine (53d)

9-Bromo-*N*-(*tert*-butyl)-1′-(methanesulfonyl)-2*H*-spiro(benzo[e]pyrimido[1,2-c][1,3]thiazine-3,4′-piperidin)-6(4*H*)-imine (52d). To the solution of 9-bromo-N-(tert-butyl)-1′-(4-methoxybenzyl)-2H-spiro(benzo[e]pyrimido[1,2-c][1,3]thiazine-3,4′-piperidin)-6(4H)-imine **52a** (54.2 mg, 0.10 mmol) in CH$_2$Cl$_2$ (0.5 mL) were added Et$_3$N (28.9 μL, 0.20 mmol) and 1-chloroethyl chloroformate (21.8 μL, 0.20 mmol) at 0 °C under an Ar atmosphere. After being stirred at same temperature for 30 min, the reaction mixture was concentrated. The residue was dissolved in MeOH (2.0 mL). After being stirred under reflux for 10 min, the reaction mixture was concentrated. The residue was dissolved in CHCl$_3$, and was washed with sat. NaHCO$_3$, brine, and dried over MgSO$_4$. After concentration, the residue was dissolved in CH$_2$Cl$_2$ (1.0 mL) and Et$_3$N (28.9 μL, 0.20 mmol), and MsCl (15.5 μL, 0.20 mmol) was added at rt under an Ar atmosphere. After being stirred at rt for 10 min, the reaction mixture was washed with sat. NaHCO$_3$, brine, and dried over MgSO$_4$. After concentration, the residue was purified by flash chromatography over aluminum oxide with n-hexane–EtOAc (6:4) to give the title compound **52d** as a colorless solid (40.9 mg, 82 %): mp 177 °C (from CHCl$_3$–n-hexane); IR (neat) cm^{-1}: 1577 (C=N), 1331 (NSO$_2$), 1155 (NSO$_2$); ^{1}H-NMR (400 MHz, CDCl$_3$) δ: 1.38 (s, 9H, 3 × CH$_3$), 1.62 (t, J = 5.5 Hz, 4H, 2 × CH$_2$), 2.80 (s, 3H, CH$_3$), 3.21–3.27 (m, 2H, CH$_2$), 3.31–3.37 (m, 2H, CH$_2$), 3.46 (s, 2H, CH$_2$), 3.84 (s, 2H, CH$_2$), 7.29–7.33 (m, 2H, Ar), 8.05 (d, J = 8.5 Hz, 1H, Ar); ^{13}C-NMR (100 MHz, CDCl$_3$) δ: 29.8, 29.9 (3C), 32.0 (2C), 34.7, 42.0 (2C), 50.1, 54.4, 55.1, 124.8, 125.9, 126.9, 129.4, 130.0, 130.8, 137.9, 146.3; HRMS (FAB): m/z calcd for C$_{20}$H$_{28}$BrN$_4$O$_2$S$_2$ [M + H]$^+$ 499.0837; found: 499.0840.

Compound 53d. Using the general procedure as described for **25a**, compound **52d** (9.2 mg, 0.018 mmol) was allowed to react for 2 h with TFA (1.0 mL) and MS4Å (150 mg). Purification by flash chromatography over aluminum oxide with n-hexane–EtOAc (1:1) gave the title compound **53d** as colorless solid (5.3 mg, 65 %): mp 171–172 °C (from CHCl$_3$–n-hexane); IR (neat) cm^{-1}: 1621 (C=N), 1564 (C=N), 1320 (NSO$_2$), 1152 (NSO$_2$); ^{1}H-NMR (400 MHz, CDCl$_3$-CD$_3$OD) δ: 1.64–1.67 (m, 4H, 2 × CH$_2$), 2.82 (s, 3H, CH$_3$), 3.19–3.25 (m, 2H, CH$_2$), 3.35–3.41 (m, 2H, CH$_2$), 3.52 (s, 2H, CH$_2$), 3.92 (s, 2H, CH$_2$), 7.24 (d, J = 2.0 Hz, 1H, Ar), 7.37 (dd, J = 8.5, 2.0 Hz, 1H, Ar), 8.07 (d, J = 8.5 Hz, 1H, Ar); ^{13}C-NMR (100 MHz, CDCl$_3$-CD$_3$OD) δ: 29.0, 31.9 (2C), 34.7, 41.7 (2C), 49.2, 54.6, 124.7, 125.4, 126.0, 129.7, 130.3, 130.4, 145.3, 153.1; HRMS (FAB): m/z calcd for C$_{16}$H$_{20}$BrN$_4$O$_2$S$_2$ [M + H]$^+$ 443.0211; found: 443.0210.

3.1.88 Synthesis of 1′-(Aminocarbonyl)-9-bromo-2H-spiro(benzo[e]pyrimido[1,2-c][1,3]thiazine-3,4′-piperidin)-6(4H)-imine (53e)

1′-(Aminocarbonyl)-9-bromo-N-(tert-butyl)-2H-spiro(benzo[e]pyrimido[1,2-c][1,3]thiazine-3,4′-piperidin)-6(4H)-imine (52e). Using the general procedure as described for **52d**, 9-bromo-N-(tert-butyl)-1′-(4-methoxybenzyl)-2H-spiro (benzo[e]pyrimido[1,2-c][1,3]thiazine-3,4′-piperidin)-6(4H)-imine **52a** (54.2 mg, 0.10 mmol) was allowed to react with 1-chloroethyl chloroformate (21.8 µL, 0.20 mmol) followed with N-trimethylsilylisocyanate (26.5 µL, 0.20 mmol). Purification by flash chromatography over aluminum oxide with EtOAc–MeOH (1:0 to 9:1) gave the title compound **52e** as colorless solid (11.8 mg, 29 %): mp 203–205 °C (from CHCl$_3$–n-hexane); IR (neat) cm^{-1}: 1649 (C=O), 1577 (C=N); ^{1}H-NMR (400 MHz, CDCl$_3$) δ: 1.37 (s, 9H, 3 × CH$_3$), 1.51 (t, J = 5.6 Hz, 4H, 2 × CH$_2$), 3.37–3.51 (m, 6H, 3 × CH$_2$), 3.82 (s, 2H, CH$_2$), 4.46 (s, 2H, NH$_2$), 7.30–7.33 (m, 2H, Ar), 8.05 (d, J = 8.5 Hz, 1H, Ar); ^{13}C-NMR (100 MHz, CDCl$_3$) δ: 29.9 (3C), 30.0, 32.1 (2C), 40.3 (2C), 50.9, 54.4, 55.1, 124.7, 126.0, 126.8, 129.3, 130.0, 130.9, 137.6, 146.3, 158.0; HRMS (FAB): m/z calcd for C$_{20}$H$_{27}$BrN$_5$OS [M + H]$^+$ 464.1120; found: 464.1122.

Compound 53e. Using the general procedure as described for **25a**, compound **52e** (5.1 mg, 0.011 mmol) was allowed to react for 2 h with TFA (1.0 mL) and

MS4Å (150 mg). Purification by flash chromatography over aluminum oxide with EtOAc–MeOH (1:0 to 9:1) gave the title compound **53e** as colorless solid (4.2 mg, 94 %): mp 222 °C (from MeOH–CHCl$_3$–n-hexane); IR (neat) cm^{-1}: 1651 (C=O), 1624 (C=N), 1585 (C=N); ^{1}H-NMR (400 MHz, DMSO-d$_6$) δ: 1.31 (t, J = 5.6 Hz, 4H, 2 × CH$_2$), 3.20–3.38 (m, 4H, 2 × CH$_2$), 3.45 (s, 2H, CH$_2$), 3.84 (s, 2H, CH$_2$), 5.88 (s, 2H, NH$_2$), 7.42 (dd, J = 8.5, 1.5 Hz, 1H), 7.59 (d, J = 1.5 Hz, 1H), 8.08 (d, J = 8.5 Hz, 1H, Ar), 8.91 (s, 1H, NH). ^{13}C-NMR (125 MHz, DMSO-d$_6$) δ: 29.0, 31.6 (2C), 39.5 ± 1.0 (2C), 48.9, 54.0, 124.3, 124.7, 126.0, 129.0, 130.2, 131.2, 144.0, 149.2, 158.0; HRMS (FAB): m/z calcd for C$_{16}$H$_{19}$BrN$_5$OS [M + H]$^+$ 408.0494; found: 408.0496.

3.1.89 Determination of Anti-HIV Activity

The sensitivity of HIV-1$_{IIIB}$ strain was determined by the MAGI assay. The target cells (HeLa-CD4/CCR5-LTR/β-gal; 10^4 cells/well) were plated in 96-well flat microtiter culture plates. On the following day, the cells were inoculated with the HIV-1 (60 MAGI U/well, giving 60 blue cells after 48 h of incubation) and cultured in the presence of various concentrations of the test compounds in fresh medium. Forty-eight hours after viral exposure, all the blue cells stained with X-Gal (5-bromo-4-chloro-3-indolyl-β-D-galactopyranoside) were counted in each well. The activity of test compounds was determined as the concentration that blocked HIV-1 infection by 50 % (50 % effective concentration [EC$_{50}$]). EC$_{50}$ was determined by using the following formula:

$$EC_{50} = 10^{\wedge}[\log(A/B) \times (50 - C)/(D - C) + \log(B)],$$
wherein

A: of the two points on the graph which bracket 50 % inhibition, the higher concentration of the test compound,
B: of the two points on the graph which bracket 50 % inhibition, the lower concentration of the test compound,
C: inhibitory activity (%) at the concentration B,
D: inhibitory activity (%) at the concentration A.

References

1. Chockalingam, K., Simeon, R.L., Rice, C.M., Chen, Z.: Proc. Natl. Acad. Sci. U.S.A. **107**, 3764–3769 (2010)
2. Chamoun, A.M., Chockalingam, K., Bobardt, M., Simeon, R., Chang, J., Gallay, P., Chen, Z.: Antimicrob. Agents Chemother. **56**, 672–681 (2012)
3. Ishihara, M., Togo, H.: Tetrahedron **63**, 1474–1480 (2007)

4. Wolfe, J.P., Wagaw, S., Marcoux, J.-F., Buchwald, S.L.: Acc. Chem. Res. **31**, 805–818 (1998)
5. Hartwig, J.P.: Angew. Chem. Int. Ed. **37**, 2046–2065 (1998)
6. Kunz, K., Scholz, U., Ganzer, D.: Synlett. 2428–2439 (2003)
7. Arvela, R.K., Pasquini, S., Larhed, M.J.: Org. Chem. **72**, 6390–6396 (2007)
8. Miyaura, N., Suzuki, A.: Chem. Rev. **95**, 2457–2483 (1995)
9. Mitschb, A., Altenkämpera, M., Sattlerc, I., Schlitzer, M.: Arch. Pharm. Chem. Life Sci. **338**, 9–17 (2005)
10. Watanabe, K., Negi, S., Sugiura, Y., Kiriyama, A., Honbo, A., Iga, K., Kodama, E.N., Naitoh, T., Matsuoka, M., Kano, K.: Chem. Asian J. **5**, 825–834 (2010)
11. Drake, R.R., Neamati, N., Hong, H., Pilon, A.A., Sunthankar, P., Hume, S.D., Milne, G.W.A., Pommier, Y.: Proc. Natl. Acad. Sci. U.S.A. **95**, 4170–4175 (1998)
12. Lin, W., Li, K., Doughty, M.B.: Bioorg. Med. Chem. **10**, 4131–4141 (2002)
13. Al-Mawsawi, L.Q., Fikkert, V., Dayam, R., Witvrouw, M., Burke Jr, T.R., Borchers, C.H., Neamati, N.: Proc. Natl. Acad. Sci. U.S.A. **103**, 10080–10085 (2006)
14. Baba, M., Scgols, D., Pauwels, R., Nakashima, H., De Clercq, E.J.: Acquir. Immune. Defic. Syndr. **3**, 493–499 (1990)
15. Kilby, J.M., Eron, J.J.N.: Engl. J. Med. **348**, 2228–2238 (2003)
16. Lalezari, J.P., Henry, K., O'Hearn, M., Montaner, J.S., Piliero, P.J., Trottier, B., Walmsley, S., Cohen, C., Kuritzkes, D.R., Eron Jr, J.J., Chung, J., DeMasi, R., Donatacci, L., Drobnes, C., Delehanty, J., Salgo, M.N.: Engl. J. Med. **348**, 2175–2185 (2003)
17. Matthews, T., Salgo, M., Greenberg, M., Chung, J., DeMasi, R., Bolognesi, D.: Nat. Rev. Drug Discov. **3**, 215–225 (2004)
18. Fischl, M.A., Richman, D.D., Grieco, M.H.N.: Engl. J. Med. **317**, 185–191 (1987)
19. Merluzzi, V.J., Hargrave, K.D., Labadia, M., Grozinger, K., Skoog, M., Wu, J.C., Shih, C.-K., Eckner, K., Hattox, S., Adams, J., Rosehthal, A.S., Faanes, R., Eckner, R.J., Koup, R.A., Sullivan, J.L.: Science **250**, 1411–1413 (1990)
20. Steigbigel, R.T., Cooper, D.A., Kumar, P.N., Eron, J.E., Schechter, M., Markowitz, M., Loutfy, M.R., Lennox, J.L., Gatell, J.M., Rockstroh, J.K., Katlama, C., Yeni, P., Lazzarin, A., Clotet, B., Zhao, J., Chen, J., Ryan, D.M., Rhodes, R.R., Killar, J.A., Gilde, L.R., Strohmaier, K.M., Meibohm, A.R., Miller, M.D., Hazuda, D.J., Nessly, M.L., DiNubile, M.J., Isaacs, R.D., Nguyen, B.-Y., Teppler, H.N.: Engl. J Med. **359**, 339–354 (2008)
21. Witvrouw, M., Pannecouque, C., Switzer, W.M., Folks, T.M., De Clercq, E., Heneine, W.: Antivir. Ther. **9**, 57–65 (2004)

Chapter 4
Design and Synthesis of Photoaffinity Probes and Their Application to Target Identification Study of PD 404182

The comparative time of drug addition study using standard anti-HIV agents demonstrated that PD 404182 **1** showed a similar antiviral profile against HIV-1$_{IIIB}$ infection with that of DS 5000 (adsorption inhibitor) and enfuvirtide (fusion inhibitor). This suggests that compound **1** apparently impairs virus replication at the early stage of HIV infection. Additionally, the antiviral activities of **1** against multiple HIV clades suggest that the target molecule of **1** is not chemokine receptors (CC chemokine receptor type 5 or CXC chemokine receptor type 4). However, the mode of action and mechanism of antiviral activity of **1** were not fully elucidated.

Photoaffinity labeling is an efficient approach to identify the target protein(s) of biologically active molecules [1–4]. In modern drug discovery, there have been a number of successful examples that have determined the target molecules and identified the binding site through the formation of a covalent bond between the ligand and the specific protein [5–7]. In general, photoaffinity probes contain three functional groups: a bioactive scaffold, a photoreactive group and an indicator group. A biotin-tag is widely employed as an indicator because biotinylated proteins can be detected and isolated by several immunological methods or through a biotin-avidin interaction [8–11]. A terminal alkyne is an alternative indicator for Huisgen cycloaddition-mediated conjugation with various azide-modified reporters, such as fluorescent-azide and biotin-azide after the crosslinking reaction onto the target protein(s). (For examples of alkyne-conjugated photoaffinity probes with benzophenone, see [12–16]).

Trifunctional probes for the target protein(s) of **1** and the derivatives were designed on the basis of the SAR investigations. In the SAR study, the introduction of a hydrophobic group on the benzene ring and the cyclic amidine substructures effectively improved antiviral activity (compounds **2–4**, Fig. 4.1). The author expected that these moieties would potentially take part in a favorable interaction(s) with the target molecule(s), and the incorporation of a hydrophobic and photoreactive benzophenone group on the pyrimidobenzothiazine scaffold would be tolerated. Additionally, the N-alkoxycarbonyl piperidine group onto the amidine substructure of **1** reproduced potent anti-HIV activity (compound **5**),

T. Mizuhara, *Development of Novel Anti-HIV Pyrimidobenzothiazine Derivatives*, Springer Theses, DOI: 10.1007/978-4-431-54445-6_4, © Springer Japan 2013

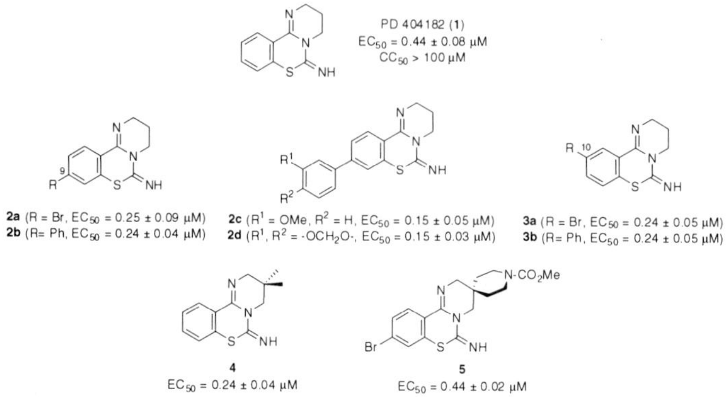

Fig. 4.1 Structures and anti-HIV activity of PD 404182 and the derivatives **2–5**

indicating that this part could be used as a linkage position for the addition of functional groups.

With this in mind, the author designed three photoaffinity probes. Compound **6** is modified with indicator biotin via a photoreactive benzophenone group onto the benzene ring substructure (Fig. 4.2). Compound **7** equips the biotin and benzophenone groups on the right-part amidine moiety. The biotin moiety is conjugated with benzophenone via a polyethylene glycol (PEG) linker as the spacer. Compound **8** is an alkyne-containing derivative.

Synthesis of the probe **6** started with the preparation of benzophenone boronic acid pinacol ester **11** (Scheme 4.1). Condensation of *p*-(hydroxymethyl)benzoic acid **9** and *N*,*O*-dimethyl-hydroxylamine followed by TBDPS protection of a primary hydroxyl group gave an amide **10**. Subsequent nucleophilic addition of an

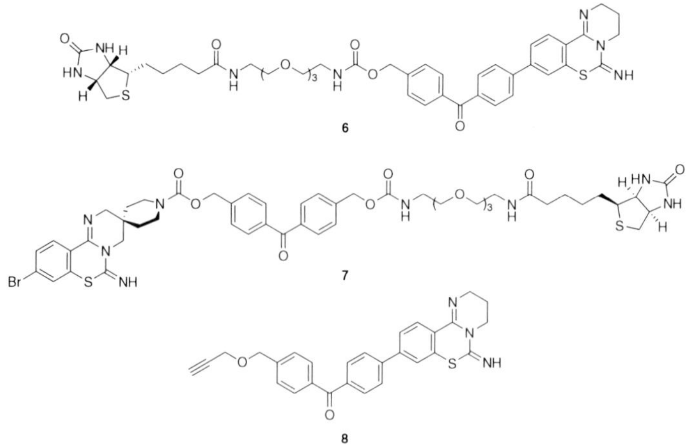

Fig. 4.2 Structures of photoaffinity probes **6–8**

in situ-generated organolithium compound easily provided the desired boronate **11** [17]. Alkylation of compound **2a** with *p*-methoxybenzyl (PMB) bromide followed by Suzuki–Miyaura cross coupling with compound **11** afforded a benzophenone-conjugated pyrimidobenzothiazine **13**. Desilylation of **13** and the subsequent reaction with *p*-nitrophenyl chloroformate afforded the carbonate **16**. The biotin moiety was incorporated by reaction of **16** with biotin-PEG-NH$_2$ (**15**) which was prepared by catalytic hydrogenation of azide **14** [18]. TFA-mediated deprotection of the PMB group in compound **17** provided the desired probe **6**.

Synthesis of the biotin-conjugated probe **7** is outlined in Scheme 4.2. PMB protection of compound **18** followed by the selective removal of the PMB group on the piperidine ring provided compound **20**. Separately, the synthesis of biotin-benzophenone adduct **23** started from 4-(*tert*-butyldiphenylsilyloxy)methyl-4'-(hydroxymethyl)benzophenone **21** [19]. The treatment of **21** with chloroformate furnished a carbonate **22**. Biotin-PEG-NH$_2$ **15** was successfully conjugated onto **22** to give the biotin-benzophenone adduct **23**. Desilylation of **23**, treatment with *p*-nitrophenyl chloroformate and coupling with **20** provided a biotin/benzophenone-conjugated **26**. PMB deprotection of **26** afforded the desired probe **7**.

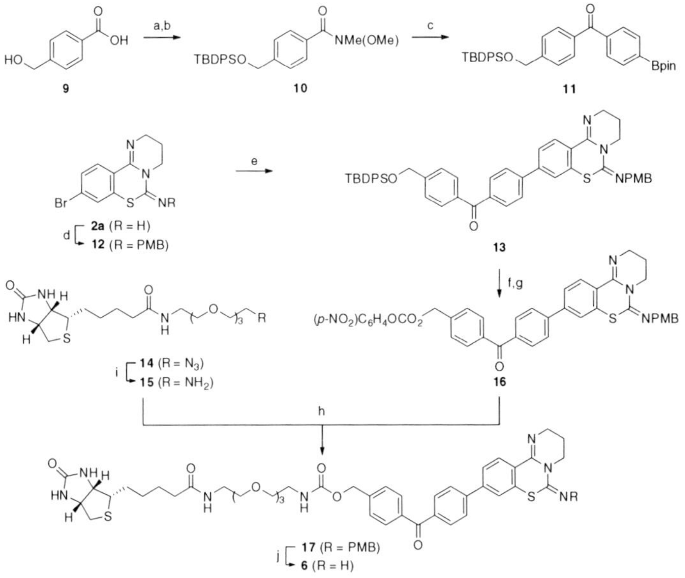

Scheme 4.1 Synthesis of biotin-conjugated probe **6** *Reagents and conditions.* (**a**) HNMe(OMe)·HCl, EDC·HCl, HOBt·H$_2$O, Et$_3$N, DMF, rt; (**b**) TBDPSCl, Et$_3$N, DMAP, CH$_2$Cl$_2$, rt, 49 % [2 steps (**a, b**)]; (*c*) 2-(4-bromophenyl)-4,4,5,5-tetramethyl-1,3,2-dioxaborolane, *t*-BuLi, −78 °C to rt, THF, 83 %; (**d**) *t*-BuOK, DMF, 0 °C, then PMBBr, rt, 98 %; (**e**) **11**, Pd(PPh$_3$)$_4$, PdCl$_2$(dppf)·CH$_2$Cl$_2$, K$_2$CO$_3$, toluene, EtOH, H$_2$O, reflux, 96 %; (**f**) TBAF, THF, rt; (**g**) *p*-nitrophenyl chloroformate, pyridine, CH$_2$Cl$_2$, reflux; (**h**) Et$_3$N, DMF, rt to 40 °C, 46 % [3 steps (**f, g, h**)]; (**i**) H$_2$, 10 % Pd/C, MeOH, rt; (**j**) MS4Å, TFA, CHCl$_3$, rt, 35 %

The author next investigated the synthesis of alkyne-containing probe **8** (Scheme 4.3). Suzuki–Miyaura cross coupling of compound **27** with boronate **11** gave compound **28**. Subsequent modifications including desilylation, propargylation, and removal of the *tert*-butyl group provided the expected alkyne-conjugated probe **8**.

The antiviral activities of probes **6–8** against HIV-1$_{IIIB}$ were measured by multinuclear activation of a galactosidase indicator (MAGI) assay. Both biotin-conjugated probes **6** and **7** showed potent anti-HIV activity with EC$_{50}$ values of 6.87 and 5.11 μM, respectively (Table 4.1). These activities were slightly lower than that of compound **1**; however, the incorporation of large functional groups including benzophenone, the PEG linker and the biotinyl reporter was largely tolerated. Alkyne-conjugated probe **8** potently inhibited HIV infection (EC$_{50}$ = 0.64 μM). These probes **6–8** represent promising tools for the identification of the target molecule(s) of compound **1** and the derivatives.

Probes **6** and **7** were applied to the experiment for target identification of compound **1** and the derivatives. After HIV-1-infected H9 cells (H9IIIB) were

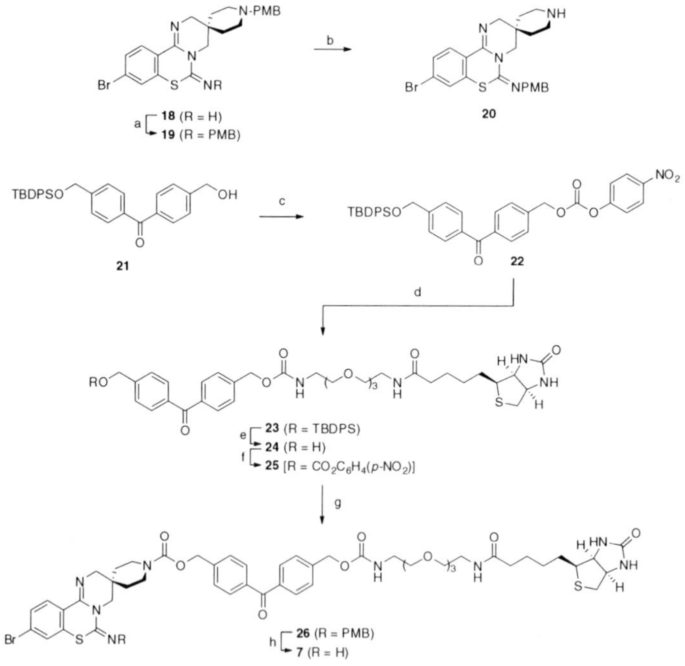

Scheme 4.2 Synthesis of biotin-conjugated probe **7**. *Reagents and conditions*: (**a**) *t*-BuOK, DMF, 0 °C, then PMBBr, rt, 81 %; (**b**) 1-chloroethylchloroformate, Et$_3$N, CH$_2$Cl$_2$, 0 °C, then MeOH, reflux; (**c**) 4-nitrophenylchloroformate, pyridine, CH$_2$Cl$_2$, reflux; (**d**) **15**, Et$_3$N, DMF, rt, quant. [2 steps (**c**, **d**)]; (**e**) HF-pyridine, THF, 0 °C to rt, 73 %; (**f**) 4-nitrophenylchloroformate, pyridine, CH$_2$Cl$_2$, reflux, 80 %; (**g**) **20**, Et$_3$N, DMF, rt; (**h**) MS4Å, TFA, CHCl$_3$, rt, 36 % [2 steps (**g**, **h**)]

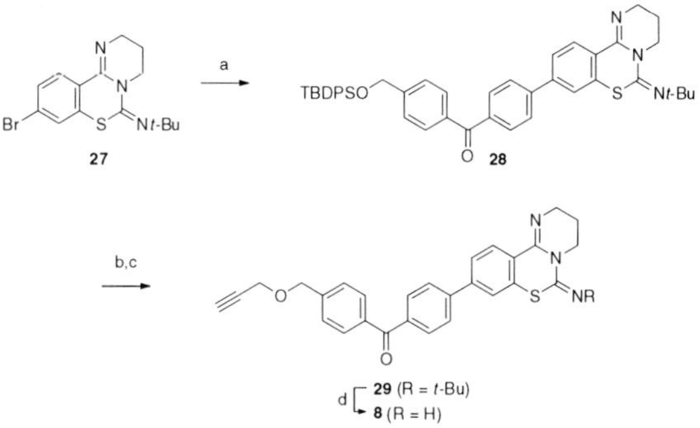

Scheme 4.3 Synthesis of alkyne-conjugated probe **8**. *Reagents and conditions:* (**a**) **11**, Pd(PPh$_3$)$_4$, PdCl$_2$(dppf)·CH$_2$Cl$_2$, K$_2$CO$_3$, toluene, EtOH, H$_2$O, reflux, 71 %; (**b**) TBAF, THF, rt; (**c**) NaH, THF, propargyl bromide, 0 °C to rt, 60 % [2 steps (**b**, **c**)]; (**d**) MS4Å, TFA, CHCl$_3$, reflux, 92 %

Table 4.1 Anti-HIV activities of the probes **6–8**

Compound	EC$_{50}$ (µM)^a
PD 404182 (**1**)	0.44 ± 0.08
6	6.87 ± 2.22
7	5.11 ± 1.31
8	0.64 ± 0.06

^{a} EC$_{50}$ values represent the concentration of compound required to inhibit the HIV-1 infection by 50 % and were obtained from three independent experiments

incubated with a probe (**6** in the presence or absence of **3a**, or **7**) for 1 h, the cells were exposed to UV–Vis light (>300 nm) for 1 min. After cell lysis, the biotinylated proteins were captured with NeutrAvidin agarose beads. The whole was subjected to separation by SDS-PAGE followed by Western blot analysis.

Eight bands of 95, 80, 75, 70, 60, 55, 48 and 40 kDa proteins were observed from the cell samples incubated with probe **6** (Lane A, Fig. 4.3). These bands were competed by unlabeled compound **3a**, suggesting that the labeling was PD 404182-specific (Lane C). In contrast, these bands, with the exception of the 70 and 40 kDa bands, were not detected in the cells incubated with probe **7** (Lane B). This observation indicated that the potential target proteins did not fully interact with the benzophenone group on the right-part amidine moiety in the pyrimidobenzothiazine scaffold of **7**. This experiment demonstrated that the synthesized probe **6** could be useful for the identification of the target protein(s) of compound **1**.

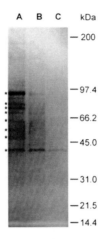

Fig. 4.3 Western Blot Analysis of the Photolabeled Proteins with Biotin-Conjugated Probes **6** and **7**; H9IIIB cells were incubated with (**A**) 20 μM probe **6**, (**B**) 20 μM probe **7**, and (**C**) 20 μM probe **6** and 40 μM compound **3a**. The cells were exposed to UV light for 1 min and were lysed. The resulting photolabeled proteins were captured onto NeutrAvidin-agarose and the whole was subjected to SDS-PAGE. The resulting gel was analyzed by Western blotting with streptavidin-HRP

In conclusion, the author designed and synthesized novel photoaffinity probes of PD 404182 with photoreactive benzophenone, and biotin or alkyne indicators. The probes exhibited equipotent or slightly less potent anti-HIV activities when compared with the activity of the parent compound **1**. Photoaffinity labeling experiments suggest that these probes could be useful in the identification of a potential target protein(s), the binding site on the target protein(s) and the mechanism(s) of action of PD 404182 derivatives.

4.1 Experimental Section

4.1.1 General Methods

All moisture-sensitive reactions were performed using syringe-septum cap techniques under an Ar atmosphere and all glasswares were dried in an oven at 80 °C for 2 h prior to use. Melting points were measured by a hot stage melting point apparatus (uncorrected). For flash chromatography, Wakogel C-300E (Wako) or aluminum oxide 90 standardized (Merck) was employed. For preparative TLC, TLC silica gel 60 F254 (Merck) or TLC aluminum oxide 60 F254 basic (Merck) were employed. For analytical HPLC, a COSMOSIL 5C18-ARII column (4.6 × 250 mm, Nacalai Tesque, Inc., Kyoto, Japan) was employed with a linear gradient of CH_3CN containing 0.1 % (v/v) NH_3 at a flow rate of 1 mL/min on a

Shimadzu LC-10ADvp (Shimadzu Corp., Ltd., Kyoto, Japan), and eluting products were detected by UV at 254 nm. ^{1}H-NMR spectra were recorded using a JEOL AL-400 or a JEOL ECA-500 spectrometer, and chemical shifts are reported in δ (ppm) relative to Me$_4$Si (CDCl$_3$) or DMSO (DMSO-d_6) as internal standards. ^{13}C-NMR spectra were recorded using a JEOL AL-400 or JEOL ECA-500 spectrometer and referenced to the residual solvent signal. ^{19}F-NMR spectra were recorded using a JEOL ECA-500 and referenced to the internal CFCl$_3$ (δ $_F$ 0.00 ppm). ^{1}H-NMR spectra are tabulated as follows: chemical shift, multiplicity (b = broad, s = singlet, d = doublet, t = triplet, q = quartet, m = multiplet), coupling constant(s), and number of protons. Exact mass (HRMS) spectra were recorded on a JMS-HX/HX 110 A mass spectrometer. Infrared (IR) spectra were obtained on a JASCO FT/IR-4100 FT-IR spectrometer with JASCO ATR PRO410-S. The purity of the probes **6–8** was determined by HPLC analysis as >95 %. Synthesis and characterization data of compounds **2a**, **18**, and **27** are shown in Chap. 3.

4.1.2 4-[(tert-Butyldiphenylsilyloxy)methyl]-N-methoxy- N-methylbenzamide (10)

To a mixture of 4-(hydroxyl- methyl)benzoic acid **9** (4.6 g, 30.0 mmol), N,O-dimethylhydroxylamine hydrochloride (14.6 g, 150.0 mmol), Et$_3$N (21.7 mL, 150.0 mmol) in DMF (300 mL) were added EDC·HCl (11.5 g, 60.0 mmol) and HOBt·H$_2$O (9.2 g, 60.0 mmol). After being stirred at rt overnight, solvent was evaporated. The residue was dissolved in EtOAc, and washed with 1 N HCl, sat. NaHCO$_3$, brine, and dried over MgSO$_4$. The filtrate was concentrated to give crude Weinreb amide (4.05 g, ca. 20.7 mmol). To the mixture of the Weinreb amide, a solution of Et$_3$N (8.98 mL, 62.1 mmol) and DMAP (252.9 mg, 2.1 mmol) in CH$_2$Cl$_2$ (138 mL) was slowly added TBDPSCl (5.83 mL, 22.8 mmol). After being stirred at rt for 3 h, the reaction mixture was quenched with water. After concentration, the residue was dissolved in EtOAc. The mixture was washed with sat. NaHCO$_3$, brine, and dried over MgSO$_4$. After concentration, the residue was purified by flash column chromatography over silica gel with n-hexane–EtOAc (3:1) to give the title compound **10** as colorless oil (6.98 g, 49 %): IR (neat) cm^{-1}: 1644 (C=O); ^{1}H-NMR (400 MHz, CDCl$_3$) δ: 1.10 (s, 9H, 3 × CH$_3$), 3.36 (s, 3H, CH$_3$), 3.57 (s, 3H, CH$_3$), 4.80 (s, 2H, CH$_2$), 7.36–7.43 (m, 8H, Ar), 7.65–7.70 (m, 6H, Ar); ^{13}C-NMR (100 MHz, CDCl$_3$) δ: 19.3, 26.8 (3C), 33.8, 61.0, 65.2, 125.4 (2C), 127.7 (4C), 128.2 (2C), 129.8 (2C), 132.6, 133.3 (2C), 135.5 (4C), 143.8, 169.9; HRMS (FAB): m/z calcd for C$_{26}$H$_{32}$NO$_3$Si [M + H]$^+$ 434.2152; found: 434.2160.

4.1.3 4-[(tert-Butyldiphenylsilyloxy)methyl]-4'-(4,4, 5,5-tetramethyl-1,3,2-dioxaborolan-2-yl) benzophenone (11)

To a solution of 1,4-dibromobenzene (3.13 g, 13.3 mmol) and 2-isopropoxy-4,4,5,5-tetramethyl-1,3,2-dioxaborolane (2.80 mL, 13.8 mmol) in anhydrous THF (60 mL) was added t-BuLi (19.4 mL, 1.55 M in pentane, 30.0 mmol) dropwise over 3 min at -78 °C under an Ar atmosphere. After being stirred at -78 °C for 30 min, additional t-BuLi (19.4 mL, 1.55 M in pentane, 30.0 mmol) was added dropwise over 3 min. After being stirred at the same temperature for additional 20 min, compound 10 (3.25 g, 7.5 mmol) was added. The reaction mixture was warmed to rt over 1 h and quenched with sat. NH$_4$Cl. The whole was extracted with EtOAc and the extract was dried over MgSO$_4$. After concentration, the residue was purified by silica gel chromatography with n-hexane–EtOAc (9:1) to give the title compound 11 as yellow oil (3.60 g, 83 %): IR (neat) cm^{-1}: 1659 (C=O); ^{1}H-NMR (400 MHz, CDCl$_3$) δ: 1.11 (s, 9H, 3 × CH$_3$), 1.37 (s, 12H, 4 × CH$_3$), 4.85 (s, 2H, CH$_2$), 7.37–7.46 (m, 8H, Ar), 7.69 (d, J = 6.6 Hz, 4H, Ar), 7.75–7.80 (m, 4H, Ar), 7.92 (d, J = 8.0 Hz, 2H, Ar); ^{13}C-NMR (100 MHz, CDCl$_3$) δ: 19.3, 24.8 (4C), 26.8 (3C), 65.2, 84.2 (2C), 125.6 (2C), 127.8 (4C), 128.9 (2C), 129.8 (2C), 130.2 (2C), 133.2 (2C), 134.5 (2C), 134.8, 135.5 (4C), 136.2, 140.0, 146.0, 196.6; HRMS (FAB): m/z calcd for C$_{36}$H$_{42}$BO$_4$Si [M + H]$^+$ 577.2945; found: 577.2949.

4.1.4 9-Bromo-3,4-dihydro-N-(p-methoxybenzyl)-2H,6H-pyrimido[1,2-c][1,3]benzothiazin-6-imine (12)

To the flask containing 9-bromo-3,4-dihydro-2H,6H-pyrimido[1,2-c] [1, 3]benzothiazin-6-imine 2a (740.4 mg, 2.50 mmol) and t-BuOK (561.1 mg, 5.00 mmol) was added DMF (10.0 mL) at 0 °C under an Ar atmosphere. After being stirred at the same temperature for 30 min, PMB-Br (729.0 µL, 5.00 mmol) was added. After being stirred at rt for 1 h, the reaction mixture was quenched with H$_2$O. The whole was extracted with EtOAc, and washed with sat. NaHCO$_3$, brine, and dried over MgSO$_4$. After concentration, the residue was purified by flash column chromatography over aluminum oxide with n-hexane–EtOAc (3:1) to give the title compound 12 as pale yellow amorphous (1.02 g, 98 %): IR (neat) cm^{-1}: 1661 (C=N), 1510 (C=N); ^{1}H-NMR (400 MHz, CDCl$_3$) δ: 1.97–2.03 (m, 2H), 3.64 (t, J = 5.7 Hz, 2H, CH$_2$), 3.80–3.84 (m, 5H, OCH$_3$, CH$_2$), 4.14 (s, 2H, CH$_2$), 6.86 (d, J = 8.5 Hz, 2H, Ar), 7.21–7.27 (m, 3H, Ar), 7.38 (dd, J = 8.2, 1.8 Hz, 1H, Ar), 7.43 (d, J = 1.8 Hz, 1H, Ar); ^{13}C-NMR (100 MHz, CDCl$_3$) δ: 19.8, 38.7, 44.3,

47.7, 55.3, 111.9, 114.1 (2C), 124.8, 127.9, 129.5, 130.2, 130.3 (2C), 132.6, 133.4, 138.7, 147.6, 159.1; HRMS (FAB): m/z calcd for $C_{19}H_{19}N_3OS$ $[M + H]^+$ 416.0432; found: 416.0431.

4.1.5 9-{4-[4-(tert-Butyldiphenylsilyloxy)methyl] benzoylphenyl}-3,4-dihydro-N-(p-methoxybenzyl)- 2H,6H-pyrimido[1,2-c][1,3]benzothiazin-6-imine (13)

Pd(PPh₃)₄ (32.8 mg, 4 mol%) and PdCl₂(dppf)·CH₂Cl₂ (17.4 mg, 3 mol %) were added to a solution of **12** (296.2 mg, 0.71 mmol) and **11** (409.4 mg, 0.71 mmol) in toluene (7.1 mL)-EtOH (4.3 mL)-1 M aq. K₂CO₃ (7.1 mL). After being stirred at reflux for 1 h, the mixture was extracted with CHCl₃. The extract was dried over MgSO₄ and concentrated. The residue was purified by flash chromatography over aluminum oxide with n-hexane–EtOAc (1:0 to 9:1) to give the title compound **13** as pale yellow amorphous (536.2 mg, 96 %): IR (neat) cm⁻¹: 1658 (C=O), 1607 (C=N), 1511 (C=N); ¹H-NMR (400 MHz, CDCl₃) δ: 1.12 (s, 9H, 3 × CH₃), 2.03–2.08 (m, 2H), 3.70 (t, $J = 5.5$ Hz, 2H, CH₂), 3.77 (s, 3H, CH₃), 3.88 (t, $J = 5.9$ Hz, 2H, CH₂), 4.19 (s, 2H, CH₂), 4.86 (s, 2H, CH₂), 6.84 (d, $J = 8.5$ Hz, 2H, Ar), 7.28 (m, 1H, Ar), 7.38-7.56 (m, 14H, Ar), 7.71 (dd, $J = 7.6$, 1.2 Hz, 4H, Ar), 7.81 (d, $J = 8.0$ Hz, 2H, Ar), 7.86 (d, $J = 8.0$ Hz, 2H, Ar); ¹³C-NMR (125 MHz, CDCl₃) δ: 19.3, 19.8, 26.8 (3C), 39.0, 44.3, 47.7, 55.2, 65.1, 112.2, 113.9 (2C), 125.6 (2C), 125.7, 127.0 (2C), 127.7 (4C), 128.7, 129.5, 129.8 (2C), 130.1 (2C), 130.2, 130.3 (2C), 130.5 (2C), 133.1 (2C), 135.0, 135.5 (4C), 136.2, 136.4, 137.0, 142.1, 143.5, 146.0, 148.2, 158.9, 195.8; HRMS (FAB): m/z calcd for $C_{49}H_{48}N_3O_3SSi$ $[M + H]^+$ 786.3186; found: 786.3178.

4.1.6 N-(2-[2-{2-(2-Aminoethoxy)ethoxy}ethoxy]ethyl)-5- [(3aS,4S,6aR)-2-oxohexahydro-1H-thieno [3,4-d]imidazol-4-yl]pentanamide (15)

10 % Pd/C (wetted with ca. 55 % water, 160.0 mg) was added to the solution of biotin-amine **14**[6] (116.0 mg, 0.26 mmol) in MeOH (2.0 mL). After being stirred at rt overnight under a H₂ atmosphere, the mixture was filtered through a Celite pad and concentrated. The crude product was used for the next step without further purification.

4.1.7 4-(4-{6-[(4-Methoxybenzyl)imino]-2,3,4,6-
tetrahydrobenzo[e]pyrimido[1,2-c][1,3]thiazin-
9-yl}benzoyl)benzyl {13-oxo-17-[(3aS,4S,6aR)-
2-oxohexahydro-1H-thieno[3,4-d]imidazol-4-yl]-3,6,
9-trioxa-12-azaheptadecyl}-carbamate (17)

To a solution of **13** (157.2 mg, 0.20 mmol) in THF (2.0 mL) was added TBAF in THF (0.50 mmol). After being stirred at rt overnight, the reaction mixture was quenched with sat. NH$_4$Cl. The whole was extracted with CHCl$_3$ and dried over MgSO$_4$. After concentration, the residue was subjected to flash column chromatography over aluminum oxide with *n*-hexane–EtOAc (5:1 to 0:1) to give the desilylated compound. To a solution of the resulting compound in CH$_2$Cl$_2$ (6.0 mL) were added *p*-nitrophenyl chloroformate (60.5 mg, 0.30 mmol) and pyridine (64.6 µL, 0.8 mmol). After being stirred under reflux for 1 h, additional *p*-nitrophenyl chloroformate (12.0 mg, 0.06 mmol) was added. After being stirred under reflux for additional 30 min, the reaction mixture was washed with brine, and dried over MgSO$_4$. After concentration, the solution of resulting residue (crude **16**) in DMF (2.0 mL) was added to the solution of **15** (ca. 0.26 mmol) and Et$_3$N (86.7 µL) in DMF (3.0 mL). After being stirred at rt for 8 h, the reaction mixture was stirred at 40 °C overnight. After concentration, the residue was purified by flash column chromatography over aluminum oxide with CHCl$_3$–MeOH (1:0 to 95:5) followed by flash column chromatography over silica gel with CHCl$_3$–MeOH (1:0 to 9:1) to give the title compound **17** as pale yellow amorphous (90.6 mg, 46 %): IR (neat) cm^{-1}: 1699 (C=O), 1656 (C=O), 1607 (C=N), 1511 (C=N); ^{1}H-NMR (500 MHz, CDCl$_3$) δ: 1.39–1.45 (m, 2H, CH$_2$), 1.57–1.74 (m, 4H, 2 × CH$_2$), 2.03-2.08 (m, 2H, CH$_2$), 2.20 (t, *J* = 6.9 Hz, 2H, CH$_2$), 2.70 (d, *J* = 12.6 Hz, 1H, CH), 2.87 (dd, *J* = 12.6, 4.6 Hz, 1H, CH), 3.12 (d, *J* = 11.7, 4.6 Hz, 1H, CH), 3.40–3.43 (m, 4H, 2 × CH$_2$), 3.54–3.71 (m, 14H, 7 × CH$_2$), 3.77 (s, 3H, CH$_3$), 3.88 (t, *J* = 6.0 Hz, 2H, CH$_2$), 4.19 (s, 2H, CH$_2$), 4.26–4.29 (m, 1H, CH), 4.45-4.47 (m, 1H, CH), 5.17 (s, 1H, NH), 5.20 (s, 2H, CH$_2$), 5.65 (s, 1H, NH), 6.07 (s, 1H, NH), 6.48 (s, 1H, NH), 6.84 (d, *J* = 8.0 Hz, 2H, Ar), 7.26–7.28 (m, 2H, Ar), 7.44–7.62 (m, 7H, Ar), 7.81 (d, *J* = 8.0 Hz, 2H, Ar), 7.85 (d, *J* = 8.0 Hz, 2H, Ar); ^{13}C-NMR (125 MHz, CDCl$_3$) δ: 19.8, 25.5, 28.0, 28.1, 35.9, 39.0, 39.1, 40.4, 40.9, 44.3, 47.7, 55.2, 55.5, 60.1, 61.7, 65.8, 69.9, 69.9, 70.0, 70.2, 70.3 (2C), 112.2, 114.0 (2C), 125.7, 127.1 (2C), 127.4 (2C), 127.6, 128.6, 129.5, 130.2 (2C), 130.3 (2C), 130.6 (2C), 135.0, 136.5, 136.7, 137.1, 141.4, 142.0, 143.7, 148.2, 156.3, 158.9, 163.9, 173.2, 195.7; HRMS (FAB): *m/z* calcd for C$_{52}$H$_{62}$N$_7$O$_9$S$_2$ [M + H]$^+$ 992.4050; found: 992.4050.

4.1.8 4-[4-(2,3,4,6-Tetrahydro-6-iminobenzo[e]pyrimido[1,2-c] [1, 3]thiazin-9-yl)benzoyl]benzyl{13-oxo-17-[(3aS, 4S,6aR)-2-oxohexahydro-1H-thieno[3,4-d]imidazol-4-yl]-3,6,9-trioxa-12-azaheptadecyl}carbamate (6)

TFA (2.0 mL) was added to a mixture of **17** (62.9 mg, 0.063 mmol) in small amount of CHCl$_3$ (1 or 2 drops) and MS4Å (300 mg, powder, activated by heating with Bunsen burner). After being stirred at rt for 4 h, Et$_3$N was added dropwise to the stirring mixture at 0 °C to adjust pH to 8–9. The whole was extracted with CHCl$_3$, and washed with sat. NaHCO$_3$, brine, and dried over MgSO$_4$. After concentration, the residue was purified by flash chromatography over aluminum oxide with CHCl$_3$–MeOH (1:0 to 95:5) followed by preparative HPLC to give the title compound **6** as colorless solid (19.3 mg, 35 %): IR (neat) cm^{-1}: 1699 (C=O), 1654 (C=O), 1621 (C=O), 1601 (C=N), 1574 (C=N); ^{1}H-NMR (500 MHz, CDCl$_3$) δ: 1.39–1.44 (m, 2H, CH$_2$), 1.60–1.76 (m, 4H, 2 × CH$_2$), 1.99–2.04 (m, 2H, CH$_2$), 2.20 (t, J = 7.4 Hz, 2H, CH$_2$), 2.71 (d, J = 12.6 Hz, 1H, CH), 2.88 (dd, J = 12.6, 5.0 Hz, 1H, CH), 3.11 (d, J = 11.7, 5.0 Hz, 1H, CH), 3.40–3.43 (m, 4H, 2 × CH$_2$), 3.54–3.63 (m, 12H, 6 × CH$_2$), 3.73 (t, J = 5.4 Hz, 2H, CH$_2$), 4.06 (t, J = 6.0 Hz, 2H, CH$_2$), 4.28 (t, J = 6.0 Hz, 1H, CH), 4.47 (t, J = 6.0 Hz, 1H, CH), 5.20 (s, 2H, CH$_2$), 5.44 (s, 1H, NH), 5.73 (s, 1H, NH), 6.37 (s, 1H, NH), 6.66 (s, 1H, NH), 7.32 (s, 1H, Ar), 7.48 (d, J = 8.0 Hz, 2H, Ar), 7.52 (d, J = 8.6 Hz, 1H, Ar), 7.69 (d, J = 8.0 Hz, 2H, Ar), 7.81 (d, J = 8.0 Hz, 2H, Ar), 7.88 (d, J = 8.0 Hz, 2H, Ar), 8.36 (d, J = 8.6 Hz, 1H, Ar); ^{13}C-NMR (125 MHz, CDCl$_3$) δ: 20.8, 25.6, 28.0, 28.2, 35.9, 39.0, 40.4, 40.9, 43.9, 44.7, 51.2, 55.6, 60.1, 61.7, 65.7, 69.9, 70.0, 70.1, 70.3 (2C), 122.0, 125.2, 125.8, 126.9 (2C), 127.4 (2C), 129.6, 129.7, 130.2 (2C), 130.7 (2C), 137.0, 141.5, 142.2, 142.9, 144.8, 146.6, 152.9, 156.3, 164.1, 173.3, 195.6; HRMS (FAB): m/z calcd for C$_{44}$H$_{54}$N$_7$O$_8$S$_2$ [M + H]$^+$ 872.3475; found: 872.3481.

4.1.9 N-[9-Bromo-1'-(4-methoxybenzyl)-2H-spiro(benzo[e]pyrimido[1,2-c][1,3]thiazine-3,4'-piperidin)-6(4H)-ylidene]-1-(4-methoxyphenyl)methanamine (19)

By a procedure identical with that described for synthesis of **12** from **2a**, the 9-bromo-1'-(4-methoxybenzyl)-2H-spiro(benzo[e]pyrimido[1,2-c][1,3]thiazine-3,4'-piperidin)-6(4H)-imine **18** (274.3 mg, 0.57 mmol) was converted into **19** as colorless amorphous (275.1 mg, 81 %): IR (neat) cm^{-1}: 1668 (C=N), 1510 (C=N); ^{1}H-NMR (400 MHz, CDCl$_3$) δ: 1.61–1.64 (m, 4H, 2 × CH$_2$), 2.36–2.42 (m, 2H, CH$_2$), 2.45–2.51 (m, 2H, CH$_2$), 3.45 (s, 2H, CH$_2$), 3.47 (s, 2H, CH$_2$), 3.55 (s, 2H, CH$_2$), 3.80 (s, 3H, CH$_3$), 3.81 (s, 3H, CH$_3$), 4.12 (s, 2H, CH$_2$), 6.82-6.87 (m, 4H,

Ar), 7.19–7.23 (m, 5H, Ar), 7.38 (dd, $J = 8.2$, 1.8 Hz, 1H, Ar), 7.44 (d, $J = 2.0$ Hz, 1H, Ar); ^{13}C-NMR (100 MHz, CDCl$_3$) δ: 28.2, 32.4 (2C), 39.1, 48.7 (2C), 54.6, 55.2, 55.3, 55.4, 62.6, 111.9, 113.7 (2C), 113.9, 114.1 (2C), 124.8, 128.0, 129.7, 130.0, 130.2 (4C), 133.4, 133.4, 138.6, 147.1, 158.8, 159.1; HRMS (FAB): m/z calcd for C$_{31}$H$_{34}$BrN$_4$O$_2$S [M + H]$^+$ 605.1586; found: 605.1585.

4.1.10 N-[9-Bromo-2H-spiro(benzo[e]pyrimido[1,2-c] [1, 3]thiazine-3,4′-piperidin)-6(4H)-ylidene]-1- (4-methoxy-phenyl)methanamine (20)

To a solution of **19** (60.6 mg, 0.10 mmol) in CH$_2$Cl$_2$ (0.5 mL) were added Et$_3$N (28.9 µL, 0.20 mmol) and 1-chloroethyl chloroformate (21.8 µL, 0.20 mmol) at 0 °C under an Ar atmosphere. After being stirred at the same temperature for 30 min, the reaction mixture was concentrated. The residue was dissolved in MeOH (2.0 mL). After being stirred under reflux for 10 min, the reaction mixture was concentrated. The residue was dissolved in CHCl$_3$, and was washed with sat. NaHCO$_3$, brine, and dried over MgSO$_4$. After concentration, the crude product was used for the next step without further purification.

4.1.11 4-[4-(tert-Butyldiphenylsilyloxymethyl)benzoyl]benzyl {13-oxo-17-[(3aS,4S,6aR)-2-oxohexahydro-1H- thieno-[3,4-d]imidazol-4-yl]-3,6,9-trioxa-12- azaheptadecyl}carbamate (23)

To a solution of 4-(*tert*-butyldiphenylsilyloxy)methyl-4′-(hydroxymethyl)benzo-phenone **21**[7] (240.3 mg, 0.50 mmol) in CH$_2$Cl$_2$ (15.0 mL) were added p-nitrophenyl chloroformate (151.2 mg, 0.75 mmol) and pyridine (161.4 µL, 2.00 mmol). After being stirred under reflux for 1 h, the reaction mixture was washed with brine, and dried over MgSO$_4$. After concentration, the solution of the resulting residue in DMF (7.5 mL) was added to a mixture of **15** (ca. 0.20 mmol) and Et$_3$N (216.8 µL) in DMF (5.0 mL). After being stirred at rt overnight, the mixture was concentrated. The residue was purified by flash column chromatography over silica gel with CHCl$_3$–MeOH (1:0 to 95:5) to give the title compound **23** as colorless amorphous (471.5 mg, quant.): IR (neat) cm^{-1}: 1700 (C=O), 1656 (C=O), 1609 (C=O); ^{1}H-NMR (400 MHz, CDCl$_3$) δ: 1.12 (s, 9H, 3 × CH$_3$), 1.39–1.46 (m, 2H, CH$_2$), 1.61–1.76 (m, 4H, 2 × CH$_2$), 2.19-2.23 (m, 2H, CH$_2$), 2.69–2.76 (m, 1H, CH), 2.85–2.90 (m, 1H, CH), 3.09–3.15 (m, 1H, CH), 3.39–3.43 (m, 4H, 2 × CH$_2$), 3.54–3.66 (m, 12H, 6 × CH$_2$), 4.26–4.33 (m, 1H, CH),

4.45–4.51 (m, 1H, CH), 4.85 (s, 2H, CH$_2$), 5.19 (s, 2H, CH$_2$), 5.54 (br s, 1H, NH), 5.68 (br s, 1H, NH), 6.55 (br s, 1H, NH), 6.72 (br s, 1H, NH), 7.36-7.48 (m, 10H, Ar), 7.69 (d, $J = 7.6$ Hz, 2H, Ar), 7.70 (d, $J = 7.6$ Hz, 2H, Ar), 7.77 (d, $J = 5.5$ Hz, 2H, Ar), 7.79 (d, $J = 5.5$ Hz, 2H, Ar); ^{13}C-NMR (100 MHz, CDCl$_3$) δ: 19.3, 25.5, 26.8 (3C), 28.1, 28.2, 35.9, 39.1, 40.5, 40.9, 55.5, 60.1, 61.7, 65.1, 65.8, 69.9, 70.0, 70.0, 70.2, 70.4 (2C), 125.6 (2C), 127.4 (2C), 127.8 (4C), 129.8 (2C), 130.1 (2C), 130.2 (2C), 133.2 (2C), 135.5 (4C), 136.1, 137.4, 141.2, 146.0, 156.3 163.9, 173.2, 196.0; HRMS (FAB): m/z calcd for C$_{50}$H$_{65}$N$_4$O$_9$SSi [M + H]$^+$ 925.4242; found: 925.4246.

4.1.12 4-[4-(Hydroxymethyl)benzoyl]benzyl{13-oxo-17-[(3aS,4S,6aR)-2-oxohexahydro-1H-thieno[3,4-d]imidazol-4-yl]-3,6,9-trioxa-12-azaheptadecyl}carbamate (24)

To a solution of **23** (383.0 mg, 0.41 mmol) in THF (8.2 mL) was added HF-pyridine (617.7 µL, Aldrich) at 0 °C. After being stirred at rt overnight, the reaction was quenched with sat. NaHCO$_3$. The whole was extracted with CHCl$_3$, and washed with water and brine, and dried over MgSO$_4$. After concentration, the residue was purified by preparative TLC over silica gel with CHCl$_3$–MeOH (85:15) to give the title compound **24** as colorless oil (204.2 mg, 73 %): IR (neat) cm^{-1}: 1696 (C=O), 1650 (C=O), 1609 (C=O); ^{1}H-NMR (400 MHz, CDCl$_3$) δ: 1.34–1.41 (m, 2H, CH$_2$), 1.55–1.73 (m, 4H, 2 × CH$_2$), 2.07 (br s, 1H, OH), 2.16 (t, $J = 7.4$ Hz, 2H, CH$_2$), 2.68 (d, $J = 12.9$ Hz, 1H, CH), 2.85 (dd, $J = 12.9, 4.9$ Hz, 1H, CH), 3.08 (dd, $J = 11.8, 7.4$ Hz 1H, CH), 3.37–3.42 (m, 4H, 2 × CH$_2$), 3.51–3.64 (m, 12H, 6 × CH$_2$), 4.23 (t, $J = 6.2$ Hz, 1H, CH), 4.43 (t, $J = 6.2$ Hz, 1H, CH), 4.78 (s, 2H, CH$_2$), 5.18 (s, 2H, CH$_2$), 5.51 (br s, 1H, NH), 5.82 (br s, 1H, NH), 6.34 (br s, 1H, NH), 6.75 (br s, 1H, NH), 7.45 (d, $J = 8.3$ Hz, 2H, Ar), 7.48 (d, $J = 8.3$ Hz, 2H, Ar), 7.76 (d, $J = 8.0$ Hz, 2H, Ar), 7.77 (d, $J = 8.0$ Hz, 2H, Ar); ^{13}C-NMR (125 MHz, CDCl$_3$) δ: 25.5, 28.0, 28.2, 35.8, 39.1, 40.4, 40.9, 55.6, 60.2, 61.8, 64.2, 65.7, 69.9, 69.9 (2C), 70.1, 70.3 (2C), 126.4 (2C), 127.3 (2C), 130.2 (2C), 130.2 (2C), 136.2, 137.1, 141.3, 146.4, 156.4, 164.1, 173.5, 196.0; HRMS (FAB): m/z calcd for C$_{34}$H$_{47}$N$_4$O$_9$S [M + H]$^+$ 687.3064; found: 687.3058.

4.1.13 4-(4-{[[(4-Nitrophenoxy)carbonyloxy]methyl} benzoyl)benzyl-13-oxo-17-[(3aS,4S,6aR)-2-oxohexahydro-1H-thieno[3,4-d]imidazol-4-yl]-3,6,9-trioxa-12-azaheptadecylcarbamate (25)

To a solution of **24** (28.2 mg, 0.04 mmol) in CH$_2$Cl$_2$ (1.2 mL) were added *p*-nitrophenyl chloroformate (24.8 mg, 0.12 mmol) and pyridine (13.2 μL, 0.16 mmol). After being stirred under reflux for 1 h, the reaction mixture was washed with brine, and dried over MgSO$_4$. After concentration, the residue was purified by preparative TLC over aluminum oxide with CHCl$_3$–MeOH (9:1) to give the title compound **25** as colorless amorphous (27.9 mg, 80 %): IR (neat) cm^{-1}: 1768 (C=O), 1698 (C=O), 1656 (C=O), 1612 (C=O); ^{1}H-NMR (400 MHz, CDCl$_3$) δ: 1.38-1.45 (m, 2H, CH$_2$), 1.59-1.76 (m, 4H, 2 × CH$_2$), 2.20 (t, *J* = 7.4 Hz, 2H, CH$_2$), 2.72 (d, *J* = 12.7 Hz, 1H, CH), 2.88 (dd, *J* = 12.7, 4.9 Hz, 1H, CH), 3.12 (dd, *J* = 11.8, 7.4 Hz, 1H, CH), 3.38–3.44 (m, 4H, 2 × CH$_2$), 3.55-3.63 (m, 12H, 6 × CH$_2$), 4.28 (t, *J* = 6.0 Hz, 1H, CH), 4.47 (t, *J* = 6.0 Hz, 1H, CH), 5.19 (s, 2H, CH$_2$), 5.38 (s, 2H, CH$_2$), 5.52 (br s, 1H, NH), 5.69 (br s, 1H, NH), 6.44 (br s, 1H, NH), 6.66 (br s, 1H, NH), 7.41 (d, *J* = 9.3 Hz, 2H, Ar), 7.47 (d, *J* = 8.0 Hz, 2H, Ar), 7.56 (d, *J* = 8.0 Hz, 2H, Ar), 7.79 (d, *J* = 8.0 Hz, 2H, Ar), 7.84 (d, *J* = 8.0 Hz, 2H, Ar), 8.29 (d, *J* = 9.3 Hz, 2H, Ar); ^{13}C-NMR (CDCl$_3$, 100 MHz) δ: 25.5, 28.1, 28.2, 35.9, 39.1, 40.5, 40.9, 55.5, 60.2, 61.8, 65.8, 69.9, 70.0, 70.0 (2C), 70.2, 70.4 (2C), 121.7 (2C), 125.3 (2C), 127.5 (2C), 128.1 (2C), 130.2 (2C), 130.4 (2C), 136.8, 137.9, 138.6, 141.7, 145.5, 152.4, 155.4, 156.3, 163.9, 173.3, 195.5; HRMS (FAB): *m/z* calcd for C$_{41}$H$_{50}$N$_5$O$_{13}$S [M + H]$^+$ 852.3126; found: 852.3127.

4.1.14 4-(4-{3,17-Dioxo-21-[(3aS,4S,6aR)-2-oxohexahydro-1H-thieno[3,4-d]imidazol-4-yl]-2,7,10,13-tetraoxa-4,16-diazahenicosyl}benzoyl)benzyl-9-bromo-6-imino-4,6-dihydro-2H-spiro(benzo[e]pyrimido[1,2-c][1,3]thiazine-3,4'-piperidine)-1'-carboxylate (7)

To a solution of **20** (ca. 0.027 mmol) in DMF (0.4 mL) were added Et$_3$N (11.7 μL, 0.081 mmol) and the solution of **25** (23.3 mg, 0.027 mmol) in DMF (0.4 mL) at rt. After being stirred at the same temperature for 1 h, the reaction mixture was concentrated. The residue was subjected to preparative TLC over silica gel with CHCl$_3$–MeOH (9:1) to give crude imine **26**. By a procedure identical with that described for synthesis of **6** from **17**, the crude **26** was converted into **7** as a colorless amorphous (10.4 mg, 36 %): IR (neat) cm^{-1}: 1699 (C=O), 1655 (C=O), 1612 (C=O), 1573 (C=N); ^{1}H-NMR (400 MHz, CDCl$_3$) δ: 1.39-1.46 (m, 2H, CH$_2$), 1.53 (d, *J* = 5.6 Hz, 4H, 2 × CH$_2$), 1.61–1.72 (m, 4H, 2 × CH$_2$), 2.20 (t,

$J = 7.3$ Hz, 2H, CH$_2$), 2.71 (d, $J = 12.7$ Hz, 1H, CH), 2.89 (dd, $J = 12.7, 4.9$ Hz, 1H, CH), 3.12 (d, $J = 12.1, 7.3$ Hz, 1H, CH), 3.39-3.44 (m, 4H, 2 × CH$_2$), 3.53–3.63 (m, 18H, 9 × CH$_2$), 3.93 (s, 2H, CH$_2$), 4.28 (t, $J = 5.7$ Hz, 1H, CH), 4.47 (t, $J = 6.5$ Hz, 1H, CH), 5.14 (s, 1H, NH), 5.19 (s, 2H, CH$_2$), 5.22 (s, 2H, CH$_2$), 5.68 (s, 1H, NH), 6.01 (s, 1H, NH), 6.52 (s, 1H, NH), 7.22 (d, $J = 2.0$ Hz, 1H, Ar), 7.34 (dd, $J = 8.8, 2.0$ Hz, 1H, Ar), 7.45 (d, $J = 8.0$ Hz, 2H, Ar), 7.46 (d, $J = 8.0$ Hz, 2H, Ar), 7.79 (m, 4H, Ar), 8.10 (d, $J = 8.8$ Hz, 1H, Ar); ^{13}C-NMR (100 MHz, CDCl$_3$) δ: 25.5, 28.1, 28.1, 29.6, 32.2 (2C), 35.8, 39.1, 39.9 (2C), 40.5, 40.9, 49.9, 54.6, 55.4, 60.1, 61.8, 65.8, 66.4, 69.9, 70.0 (2C), 70.2, 70.4 (2C), 125.0, 125.3, 126.0, 127.3 (2C), 127.4 (2C), 129.6, 130.2 (2C), 130.3 (2C), 130.4, 130.6, 137.0, 137.1, 141.4, 141.5, 145.1, 152.6, 155.0, 156.3, 163.8, 173.3, 195.7; HRMS (FAB): m/z calcd for C$_{50}$H$_{62}$BrN$_8$O$_{10}$S$_2$ [M + H]$^+$ 1077.3214; found: 1077.3213.

4.1.15 N-(tert-Butyl)-9-{4-[4-(tert-butyldiphenylsilyloxy) methyl]benzoylphenyl}-3,4-dihydro-2H,6H-pyrimido[1,2-c][1,3]benzothiazin-6-imine (28)

Compound **27** (2.17 g, 6.17 mmol) was subjected to the general cross-coupling procedure as described for the synthesis of **13** to give the title compound **28** as colorless solid (3.16 g, 71 %): mp 152–153 °C (from CHCl$_3$–n-hexane): IR (neat) cm^{-1}: 1656 (C=O), 1623 (C=N), 1593 (C=N); ^{1}H-NMR (400 MHz, CDCl$_3$) δ: 1.12 (s, 9H, 3 × CH$_3$), 1.41 (s, 9H, 3 × CH$_3$), 1.91–1.97 (m, 2H), 3.65 (t, $J = 5.4$ Hz, 2H, CH$_2$), 3.90 (t, $J = 6.2$ Hz, 2H, CH$_2$), 4.86 (s, 2H, CH$_2$), 7.37–7.48 (m, 10H, Ar), 7.69–7.71 (m, 6H, Ar), 7.81 (d, $J = 8.3$ Hz, 2H, Ar), 7.88 (d, $J = 8.3$ Hz, 2H, Ar), 8.30 (d, $J = 8.5$ Hz, 1H, Ar); ^{13}C-NMR (100 MHz, CDCl$_3$) δ: 19.3, 21.9, 26.8 (3C), 30.0 (3C), 45.2, 45.5, 54.2, 65.2, 123.0, 124.9, 125.7 (2C), 126.9 (2C), 127.4, 127.8 (4C), 129.1, 129.8 (2C), 129.9, 130.2 (2C), 130.7 (2C), 133.2 (2C), 135.5 (4C), 136.2, 137.2, 138.0, 141.7, 143.2, 146.1, 147.6, 195.9; HRMS (FAB): m/z calcd for C$_{45}$H$_{48}$N$_3$O$_2$SSi [M + H]$^+$ 722.3237; found: 722.3244.

4.1.16 N-(tert-Butyl)-3,4-dihydro-9-[4-(4-propargyloxymethyl)benzoylphenyl]-2H,6H-pyrimido[1,2-c][1,3]benzo- thiazin-6-imine (29)

To a solution of **28** (200.0 mg, 0.28 mmol) in THF (2.8 mL) was added TBAF in THF (0.55 mmol). After being stirred at rt for 2 h, the reaction mixture was quenched with sat. NH$_4$Cl. The whole was extracted with EtOAc, and washed with brine, and dried over MgSO$_4$. The filtrate was concentrated. To the solution of the resulting residue in THF (2.8 mL) was added NaH (22.8 mg, 0.55 mmol, 60 % oil suspension) at 0 °C. After being stirred at the same temperature for 30 min, propargyl bromide (31.5 μL, 0.42 mmol) was added dropwise. After being stirred at rt overnight, the reaction was quenched with water. The whole was extracted with EtOAc, and washed with brine, and dried over MgSO$_4$. After concentration, the residue was purified by flash column chromatography over aluminum oxide with n-hexane–EtOAc (5:1) to give the title compound **29** as colorless solid (87.2 mg, 60 %): mp 133–135 °C (from CHCl$_3$–n-hexane): IR (neat) cm^{-1}: 1656 (C=O), 1620 (C=N), 1593 (C=N); ^{1}H-NMR (400 MHz, CDCl$_3$) δ: 1.41 (s, 9H, 3 × CH$_3$), 1.91–1.97 (m, 2H), 2.50 (t, J = 2.3 Hz, 1H, CH), 3.65 (t, J = 5.5 Hz, 2H, CH$_2$), 3.90 (t, J = 6.1 Hz, 2H, CH$_2$), 4.25 (d, J = 2.3 Hz, 2H, CH$_2$), 4.71 (s, 2H, CH$_2$), 7.39 (d, J = 1.7 Hz, 1H, Ar), 7.46–7.50 (m, 3H, Ar), 7.70 (d, J = 8.0 Hz, 2H, Ar), 7.82 (d, J = 8.0 Hz, 2H, Ar), 7.87 (d, J = 8.0 Hz, 2H, Ar), 8.30 (d, J = 8.3 Hz, 1H, Ar); ^{13}C-NMR (100 MHz, CDCl$_3$) δ: 21.9, 30.0 (3C), 45.2, 45.4, 54.2, 57.6, 70.9, 75.0, 79.3, 123.0, 124.8, 126.9 (2C), 127.3, 127.5 (2C), 129.1, 129.9, 130.2 (2C), 130.7 (2C), 136.9, 137.0, 137.9, 141.6, 142.2, 143.4, 147.5, 195.7; HRMS (FAB): m/z calcd for C$_{32}$H$_{32}$N$_3$O$_2$S [M + H]$^+$ 522.2215; found: 522.2207.

4.1.17 *3,4-Dihydro-9-[4-(4-propargyloxymethyl) benzoylphenyl]-2H,6H-pyrimido[1,2-c][1,3]benzothiazin-6-imine (8)*

Using a procedure identical with that described for synthesis of **6** from **17**, the imine **29** (42.8 mg, 0.08 mmol) was allowed to react under reflux for 1 h with TFA (2.0 mL) and MS4Å (300 mg). Purification by flash chromatography over aluminum oxide with n-hexane–EtOAc (9:1 to 1:1) gave the title compound **8** as colorless solid (35.4 mg, 92 %): mp 159–160 °C (from CHCl$_3$–n-hexane): IR (neat) cm^{-1}: 1654 (C=O), 1619 (C=N), 1573 (C=N); ^{1}H-NMR (400 MHz, CDCl$_3$) δ: 1.96–2.04 (m, 2H), 2.50 (t, J = 2.4 Hz, 1H, CH), 3.72 (t, J = 5.6 Hz, 2H, CH$_2$), 4.05 (t, J = 6.1 Hz, 2H, CH$_2$), 4.25 (d, J = 2.4 Hz, 2H, CH$_2$), 4.71 (s, 2H, CH$_2$), 7.26–7.31 (m, 2H, Ar, NH), 7.48–7.51 (m, 3H, Ar), 7.67–7.89 (m, 6H, Ar), 8.33 (d, J = 8.5 Hz, 1H, Ar); ^{13}C-NMR (100 MHz, CDCl$_3$) δ: 21.0, 43.8, 45.0, 57.6, 70.9, 75.0, 79.3, 122.0, 125.1, 126.3, 126.9 (2C), 127.5 (2C), 129.6, 129.7, 130.2 (2C), 130.7 (2C), 137.0, 137.1, 142.2, 142.3, 143.0, 146.2, 153.0, 195.7; HRMS (FAB): m/z calcd for C$_{28}$H$_{24}$N$_3$O$_2$S [M + H]$^+$ 466.1589; found: 466.1589.

4.1.18 Photoaffinity Labeling Experiments Using HIV-1-Infected H9 Cells (H9IIIB)

1 μL of probe **6** or **7** (10 mM solution in DMSO) was added to H9 cells chronically infected with HIV-1 (H9IIIB) in D-MEM with 10 % fetal bovine serum (500 μL, 0.5 × 10⁶ cells). For the competitive evaluation (Fig. 4.3, lane C), 2 μL of compound **3a** (10 mM solution in DMSO) was also added. The cells were incubated at 37 °C for 1 h. Then the cells were photolabeled by irradiation by UV (MUV-202U, Moritex Co., Japan) at rt for 1 min at a distance of 3 cm through a longpass filter (LU0300, Asahi spectra Co.). The mixture was centrifuged at 200 × g for 5 min and the supernatant was removed. The cells were washed with PBS once and were lysed in RIPA buffer containing 1 % protease inhibitor cocktail (Nacalai Tesque, Inc., Japan) at 4 °C for 30 min. After centrifugation at 16,500 × g for 15 min, the supernatant was used for the next experiment.

NeutrAvidin agarose beads (50 μL, Thermo), which were equilibrated with RIPA buffer, were treated with the supernatant containing 180 μg of proteins and were incubated at 4 °C for 1 h. The beads were then centrifuged at 9,100 × g for 30 s and washed with RIPA buffer (repeated three times). After heating the beads at 95 °C for 5 min in sample buffer [50 mM Tris–HCl (pH 8.0), 2 % SDS, 0.1 % BPB, 10 % glycerol, 2 % β-ME], the supernatants were subjected to SDS–PAGE electrophoresis (SuperSep™Ace, 5–20 %, Wako) and the separated proteins were transferred onto a PVDF membrane. The membrane was blocked with Blocking One (Nacalai Tesque, Inc.) at rt for 1 h, and was then incubated with a streptavidin–HRP conjugate (Invitrogen; 1:5,000 in PBS with 0.1 % Tween) at 4 °C overnight. The membrane was treated with Chemi-Lumi One L (Nacalai Tesque, Inc.). Biotinylated proteins were detected by Image Quant LAS 4000mini (GE Healthcare).

4.1.19 Determination of Anti-HIV Activity

The sensitivity of HIV-1$_{IIIB}$ strain was determined by the MAGI assay. The target cells (HeLa-CD4/CCR5-LTR/β-gal; 10⁴ cells/well) were plated in 96-well flat microtiter culture plates. On the following day, the cells were inoculated with the HIV-1 (60 MAGI U/well, giving 60 blue cells after 48 h of incubation) and cultured in the presence of various concentrations of the test compounds in fresh medium. Forty-eight hours after viral exposure, all the blue cells stained with X-Gal (5-bromo-4-chloro-3-indolyl- β-D-galactopyranoside) were counted in each well. The activity of test compounds was determined as the concentration that blocked HIV-1 infection by 50 % (50 % effective concentration [EC$_{50}$]). EC$_{50}$ was determined by using the following formula:

$$EC_{50} = 10^{\wedge}[\log(A/B) \times (50 - C)/(D - C) + \log(B)],$$

wherein

A   of the two points on the graph which bracket 50 % inhibition, the higher concentration of the test compound,

B   of the two points on the graph which bracket 50 % inhibition, the lower concentration of the test compound,

C   inhibitory activity (%) at the concentration B,

D   inhibitory activity (%) at the concentration A.

References

1. Dorman, G., Prestwich, G.D.: Biochemistry **33**, 5661–5673 (1994)
2. Kotzyba-Hibert, F., Kapfer, I., Goeldner, M.: Angew. Chem. Int. Ed. **34**, 1296–1312 (1995)
3. Fleming, S.A.: Tetrahedron **51**, 12479–12520 (1995)
4. Tomohiro, T., Hashimoto, M., Hatanaka, Y.: Chem. Rec. **5**, 385–395 (2005)
5. Drake, R.R., Neamati, N., Hong, H., Pilon, A.A., Sunthankar, P., Hume, S.D., Milne, G.W.A., Pommier, Y.: Proc. Natl. Acad. Sci. U.S.A. **95**, 4170–4175 (1998)
6. Lin, W., Li, K., Doughty, M.B.: Bioorg. Med. Chem. **10**, 4131–4141 (2002)
7. Al-Mawsawi, L.Q., Fikkert, V., Dayam, R., Witvrouw, M., Burke Jr, T.R., Borchers, C.H., Neamati, N.: Proc. Natl. Acad. Sci. U.S.A. **103**, 10080–10085 (2006)
8. Hofmann, K., Kiso, Y.: Proc. Natl. Acad. Sci. U.S.A. **73**, 3516–3518 (1976)
9. Hatanaka, Y., Hashimoto, M., Kanaoka, Y.: Bioorg. Med. Chem. **2**, 1367–1373 (1994)
10. Kinoshita, T., Cano-Delgado, A., Seto, H., Hiranuma, S., Fujioka, S., Yoshida, S., Chory, J.: Nature **433**, 167–171 (2005)
11. Kotake, Y., Sagane, K., Owa, T., Mimori-Kiyosue, Y., Shimizu, H., Uesugi, M., Ishihama, Y., Iwata, M., Mizui, Y.: Nat. Chem. Biol. **3**, 570–575 (2007)
12. Ballell, L., Alink, K.J., Slijper, M., Versluis, C., Liskamp, R.M., Pieters, R.J.: ChemBioChem **6**, 291–295 (2005)
13. Sieber, S.A., Niessen, S., Hoover, H.S., Cravatt, B.F.: Nat. Chem. Biol. **2**, 274–281 (2006)
14. Salisbury, C.M., Cravatt, B.F.: Proc. Natl. Acad. Sci. U.S.A. **104**, 1171–1176 (2007)
15. Kalesh, K.A., Sim, D.S., Wang, J., Liu, K., Lin, Q., Yao, S.Q.: Chem. Commun. **46**, 1118–1120 (2010)
16. Eirich, J., Orth, R., Sieber, S.A.J.: Am. Chem. Soc. **133**, 12144–12153 (2011)
17. Jiang, Q., Ryan, M., Zhichkin, P.J.: Org. Chem. **72**, 6618–6620 (2007)
18. Fusz, S., Srivatsan, S.G., Ackermann, D., Famulok, M.J.: Org. Chem. **73**, 5069–5077 (2008)
19. Denholm, A.A., George, MH, Hailes, H.C., Tiffin, P.J., Widdowson, D.A.: J. Chem. Soc., Perkin Trans. 1 **5**, 541–547 (1995)

Chapter 5
Conclusions

1. The author developed efficient methodology for the synthesis of tricyclic heterocycles related to PD 404182 based on the sp^2-carbon–heteroatom (O, N, and S) bond formations by C–H functionalization or S_NAr reaction. Starting from arene or haloarene, C–O, C–N, or C–S bonds were formed by simply changing the reactant such as nucleophiles or hetelocumulenes, respectively. These synthetic methods provide powerful approaches for the divergent preparation of pyrimido-benzoxazine, -quinazoline, or -benzo- thiazine derivatives.

2. The author has carried out the intensive SAR studies of the central 1,3-thiazin-2-imine core, the benzene part, and the cyclic amidine part of PD 404182 for the development of anti-HIV agents. The 6-6-6 fused pyrimido[1,2-c] [1,3]benzothiazine scaffold and the heteroatom arrangement in PD 404182 considerably contribute to the potent anti-HIV activity. Additionally, through optimization studies of the benzene and cyclic amidine ring parts in PD 404182, threefold more potent inhibitors were obtained compared with the lead compound. The author also revealed by a time of drug addition experiment that PD 404182 derivatives impaired HIV replication at the binding or fusion.

3. The author has developed photoaffinity probes with a photoreactive benzophenone moiety and indicator group such as biotin or alkyne for the target identification of PD 404182. By the photolabeling experiment of HIV-1-infected H9 cells using these probes, the author identified proteins specifically bound to PD 404182.

In summary, the author successfully developed novel anti-HIV pyrimidobenzothiazine derivatives. The most potent derivatives exhibited submicromolar inhibitory activity against both HIV-1 and HIV-2. These compounds could be promising agents for anti-HIV therapy because their mechanisms of action through the interaction with several proteins in viruses and/or host cells differ from that of the currently approved anti-HIV agents.

T. Mizuhara, *Development of Novel Anti-HIV Pyrimidobenzothiazine Derivatives*,
Springer Theses, DOI: 10.1007/978-4-431-54445-6_5, © Springer Japan 2013

Curriculum Vitae

Tsukasa Mizuhara

Department of Chemistry,
University of Massachusetts Amherst
710 Nt. Pleasant Street
Amherst MA 01003
E-mail: tsukasam@chem.umass.edu
https://sites.google.com/site/rotellogroup/
home

	Education
Mar. 2013	Ph. D. in Pharmaceutical Sciences
	Graduate School of Pharmaceutical Sciences, Kyoto University, Kyoto, Japan
	(Supervisor; Prof. Nobutaka Fujii)
Mar. 2010	M. Sc. in Pharmaceutical Sciences
	Graduate School of Pharmaceutical Sciences, Kyoto University, Kyoto, Japan
	(Supervisor; Prof. Nobutaka Fujii)
Mar. 2008	B. Sc. in Pharmaceutical Sciences
	Faculty of Pharmaceutical Sciences, Kobe Gakuin University, Kobe, Japan

(continued)

T. Mizuhara, *Development of Novel Anti-HIV Pyrimidobenzothiazine Derivatives*,
Springer Theses, DOI: 10.1007/978-4-431-54445-6, © Springer Japan 2013

	Research Experience
Feb. 2013-present	Postdoctoral Research (supervisor; Prof. Vincent, M. Rotello), University of Massachusetts Amherst, USA
Apr. 2008–Mar. 2013	Graduate Research (supervisor; Prof. Nobutaka Fujii), Kyoto University, Japan -"Development of the Efficient Synthetic Methods of PD 404182 and Related Tricyclic Heterocycles"- -"Structure-activity Relationship Study of Antiviral PD 404182 Derivatives"- -"Target Identification of Antiviral PD 404182 Derivatives"-
Apr. 2007–Mar. 2008	Undergraduate Research (supervisor; Prof. Munetaka Kunishima), Kobe Gakuin University, Japan -"Development of the Direct Preparation Method of Primary Amides by Reaction of Carboxylic Acids and Ammonia in Alcohols"-

	Teaching Experience
2008–2013	Teaching Assistant, Kyoto University, Japan

	Fellowships and Awards
Jan. 2013	Postdoctoral Fellowship for Research Abroad, the Japan Society for the Promotion of Science
Apr. 2010–Mar. 2013	Research Fellow (DC1), the Japan Society for the Promotion of Science
Mar. 2008	Presidential Award, Kobe Gakuin University, Kobe, Japan

Publications

8. Mizuhara, T.; Kato, T.; Hirai, A.; Kurihara, H.; Shimada, Y.; Taniguchi, M.; Maeta, H.; Togami, H.; Shimura, K.; Matsuoka, M.; Okazaki, S.; Takeuchi, T.; Ohno, H.; Oishi, S.; Fujii, N. Structure–activity relationship study of phenylpyrazole derivatives as a novel class of anti-HIV agents. *Bioorg. Med. Chem. Lett.* in press.

7. Mizuhara. T.; Oishi, S.; Ohno, H.; Shimura, K; Matsuoka, M.; Fujii, N. Design and synthesis of biotin- or alkyne-conjugated photoaffinity probes for studying the target molecules of PD 404182. *Bioorg. Med. Chem.* **2013**, *21*, 2079-2087.

6. Mizuhara. T.; Oishi, S.; Ohno, H.; Shimura, K; Matsuoka, M.; Fujii, N. Structure–activity relationship study of pyrimido[1,2-c][1,3]benzothiazin-6-imine derivatives for potent anti-HIV agents. *Bioorg. Med. Chem.* **2012**, *20*, 6334-6441.

5. Mizuhara. T.; Oishi, S.; Ohno, H.; Shimura, K; Matsuoka, M.; Fujii, N. Concise synthesis and anti-HIV activity of pyrimido[1,2-c][1,3]benzothiazin-6-imines and related tricyclic heterocycles. *Org. Biomol. Chem.* **2012**, *10*, 6792-6802.

4. Mizuhara, T. Site-selective Chemical Modification by Metallopeptide. *News and View, Chemical Biology.* **2011**, *4*, 14.

3. Mizuhara, T.; Oishi, S.; Fujii, N.; Ohno, H. Efficient Synthesis of Pyrimido[1,2-
   c][1,3]benzothiazin-6- imines and Related Tricyclic Heterocycles by S$_N$Ar-
   Type C-S, C-N, or C-O Bond Formation Using Heterocumulenes. *J. Org. Chem.*
   2010, *75*, 265–268.
2. Mizuhara, T.; Inuki, S.; Oishi, S.; Fujii, N.; Ohno, H. Cu (II)-mediated oxidative
   intermolecular *ortho* C-H functionalisation using tetrahydropyrimidine as the
   directing group. *Chem. Commun.* **2009**, 3413–3415.
1. Mizuhara, T.; Hioki, K.; Yamada, M.; Sasaki, H.; Morisaki, D.; Kunishima, M.
   Direct Preparation of Primary Amides by Reaction of Carboxylic Acids and
   Ammonia in Alcohols Using DMT-MM. *Chem. Lett.* **2008**, *37*, 1190–1191.

Patents

1. Maeta, H.; Katou, T.; Matsuoka, M.; Shimura, K.; Fujii, N.; Ohno, H.; Oishi, S.;
   Mizuhara, T. MEDICINAL AGENT CONTAINING PYRMIDOBEN
   ZOTHIAZIN-6-IMINE DERIVATIVE OR SALT THEREOF FOR
   PREVENTION AND/OR TREATMENT OF VIRAL INFECTION. PCT/
   JP2012/061890.

Printed by Publishers' Graphics LLC
DBT131010.15.15.15